RUSTIC birdhouses *and* feeders

Unique Thatched-Roof Designs Built to Bird-Friendly Specifications

Colin McGhee
North America's Master Thatcher

with Tracy Breyfogle

STOREY BOOKS

North Adams, Massachusetts

The mission of Storey Communications is to serve our customers by publishing practical information that encourages personal independence in harmony with the environment.

Edited by Deborah Balmuth, Jeff Day, and Maryann Teale Snell
Cover design by Mark Tomasi
Cover photograph by Giles Prett
Text design and production by Mark Tomasi
Production assistance by Erin Lincourt
Photographs by Giles Prett,
except for those on pages 6, 7, 12, and 99 (small inset) © Ken Wilder;
and pages vi, 3, and 4 © Joelle McGhee
Line drawings by Terry Dovastan and Associates
Indexed by Susan Olason/Indexes & Knowledge Maps

Copyright © 2000 by Storey Communications, Inc.

All rights reserved. No part of this book may be reproduced without written permission from the publisher, except by a reviewer who may quote brief passages or reproduce illustrations in a review with appropriate credits; nor may any part of this book be reproduced, stored in a retrieval system, or transmitted in any form or by any means — electronic, mechanical, photocopying, recording, or other — without written permission from the publisher.

The information in this book is true and complete to the best of our knowledge. All recommendations are made without guarantee on the part of the author or Storey Books. The author and publisher disclaim any liability in connection with the use of this information. For additional information please contact Storey Books, 210 MASS MoCA Way, North Adams, MA 01247.

Storey Books are available for special premium and promotional uses and for customized editions. For further information, please call Storey's Custom Publishing Department at (800) 793-9396.

Printed in the United States by Banta Book Group, Menasha, WI
10 9 8 7 6 5 4 3 2

Library of Congress Cataloging-in-Publication Data

McGhee, Colin, 1958 –
    Rustic birdhouses and feeders: unique thatched-roof designs built to bird-friendly specifications / Colin McGhee with Tracy Breyfogle.
        p.   cm.
    ISBN 1-58017-137-0 (alk. paper)
    1. Birdhouses—Design and construction. 2. Bird feeders—Design and construction. I. Breyfogle, Tracy, 1971. II. Title.
QL676.5 .M362000
690'.8927—dc21
                                                                                                      99-086943
                                                                                                         CIP

contents

Chapter 1:	Thatching: Not Just for the Birds	1
Chapter 2:	Getting Started: Materials, Equipment & Techniques	13
Chapter 3:	Learning to Thatch	21
Chapter 4:	Thatched Bird Feeder	31
Chapter 5:	Chickadee Box	39
Chapter 6:	Screech Owl House	49
Chapter 7:	Bat Box	61
Chapter 8:	Bluebird Box	71
Chapter 9:	Nuthatch Celtic Cottage	85
Chapter 10:	Tudor Wren House	99
Chapter 11:	Butterfly House	113
Chapter 12:	Thatched Mailbox	123
	Resources	132
	Index	134

RUSTIC
birdhouses
and feeders

thatching: not just for the birds

I will arise and go now, and go to Innisfree,
And a small cabin build there, of clay and wattles made:
Nine bean-rows will I have there, a hive for the honey-bee,
And live alone in the bee-loud glade.

William Butler Yeats

Welcome to a book that has been years, literally centuries, in the making. Although the use of thatch as a decorative touch to backyard birdhouses is a fairly recent practice, thatching itself is an ancient craft used in varying forms around the world. In addition to being readily available, the thatch material has a versatility and a beauty that have helped maintain its popularity. Even today, its appeal as an alternative building material means that the centuries-old craft of thatching a roof is getting a modern spin.

The thatching on this private estate office is made from Norfolk reed and features a decorative roof ridge pattern.

What Is Thatch?

Thatch, basically defined, is vegetation used to roof a structure. The vegetation — usually reeds or straw — is gathered, bundled, and secured to a framework on top of the building. Thatch functions both as shingle and underlayment; very rarely is there anything beyond simple framework between the thatch and the inside of the structure. For that reason, thatchers have to understand the qualities of the thatching material they use — its value as insulation and, particularly, its ability to withstand snow, wind, and rain.

The two most common types of European and American thatch are straw and reed. Both are tried-and-true choices, each bringing slightly varied qualities to the completed roof. They are not, however, the only materials suitable for thatching. Thatched roofs appear on every continent except Antarctica, and thatching materials range from plains grasses to waterproof leaves found in South American rainforests or on South Pacific islands. Early settlers to the New World used thatch as far back as 1565, but Native Americans had already been using thatch for generations. When settlers arrived in Jamestown in 1607, they found Powhatan Indians living in houses with thatched roofs. The colonists used the same thatch on their buildings.

Reed and Straw

In Europe, a roof made of reed is more durable than one made of straw. Plants such as Norfolk reed — so called because it was harvested in the Norfolk fens — were most commonly used in England, along coastal areas in the Southeast, and along the west coast of Scotland. Other reeds are harvested all over the world, commonly going by the name "water reed." The availability of reed accounts for the concentration of thatched buildings along coasts and in marshy areas.

The reed must be thoroughly dried to prevent it from shrinking once it is on the roof. Once dry, it is gathered in bundles and attached to the prepared frame of the roof. The bundles have a circumference of 24 to 27 inches, or "3 hands." Depending on the size of the project and the part of the roof, a bundle is anywhere from 3 to 7 feet long.

Straw is another common thatching material. While readily available, straw thatch does not last as long as reed. Collecting it is also more labor-intensive, since it has to be cut and thrashed before it's drawn by hand into workable bundles, known as "yearns." Because straw is so brittle, the thatcher must continually wet it while working, to keep it from breaking. The straw will shrink as it dries, so the thatcher must gauge during the process what the finished roof will look like.

A Brief History of Thatch in Europe

European thatch dates back to before the Middle Ages, when the first small, permanent villages were established. The creation of villages brought with it the

This long straw roof, which the author thatched, is on the Windmill Cottage in Essex, United Kingdom.

▶ fire in the house!

There is probably no better illustration of the fire danger posed by the prevalence of thatched roofs than London's Great Fire of 1666. When the fire first started, the city was at a loss as to how to put out the rapidly growing blaze. Buckets of water were no match for a prevailing wind that pushed the flames from rooftop to rooftop. In the end, the fire destroyed four-fifths of the city, and thatched roofs were immediately banned in London.

need for readily available, inexpensive, and durable building materials. One of these materials was thatch, an ideal roofing material that adapted well to the shift to permanent villages. People were also able to specialize in trades at this time, bringing about the rise of skilled, professional thatchers.

Gradually, these skilled craftsmen — and the homes they built — became more refined, leading to a variety of unique thatching styles. The popularity of thatch and the effort to create well-crafted roofs in this style reached a pinnacle in the late 18th century. But as with all trends, thatch was bound to be replaced over time. It was not, after all, the perfect choice for a society moving into the industrial age.

With the arrival of industry came the introduction of more effective ways of farming the land, including the new steam-powered tractor. As tractors passed through a village, they occasionally threw large cinders into the air. While thatch rarely catches fire from a stray spark, these cinders were large enough to set roofs ablaze. Add to that the fact that a fire department consisted of a line of men with buckets, and it is easy to see why something had to be done about this dangerous situation.

Changing times demanded something "different." As towns got larger, a faster-to-build and less flammable roof became essential. Thatch took a back seat to new roofing materials, such as tar, slate, copper, and lead, which eventually led to modern-day shingles.

Thatched roofs in the 20th century, however, benefit from the efforts of modern science. Roofs can now be made fire retardant, to reduce the risk of fire. This and other improvements in building technology make thatched roofs much less of a threat today than in the 18th century.

One of the author's straw thatching projects in progress.

Benefits of Thatch

Today, thatch is generally acknowledged for its visual appeal and as a way to add a historic or whimsical touch to homes. While it does impart a storybook quality to many structures, thatch still retains the practical aspects of its past. In addition to its beauty, one of its best qualities is its ability to act as an effective insulation. And believe it or not, repairing a thatched roof is much easier than you would think.

Rain Resistance

Thatch is a thick covering with thousands of pockets of air between and inside the stems of the thatching. These air pockets give it the ability to insulate a building — in both warm and cold weather.

Norfolk reed, which is commonly used on thatched roofs in the United States and Britain, offers an insulation or R-value of 40. Conventional insulation, such as fiberglass, usually rates between 30 and 50.

Upon first seeing a thatched roof, many people see a layer of unprotected reed or straw and are quick to question its ability to keep the inside of a house dry. In reality, a well-thatched roof is impenetrable to rain and pests. From the outside, it may look like a thin covering, but a closer examination reveals a layer at least a foot thick.

In addition, any building to be thatched must have a roof that is set at a minimum pitch of 45 degrees. Precipitation travels down the steep slope of the completed roof to the ground long before finding its way to rooms inside. In fact, the top inch or so of the thatch is the only part to ever feel the effect of the elements.

Ease of Repair

Although new thatch is usually 12 inches thick, a thatcher may put a layer on top of the previous layer if a thatched roof is in need of repair. The result can be an old, and very thick, layer of straw. I've stripped up to 4 feet of straw off of old roofs in England. Usually, unless the original roof was destroyed or burnt in a fire, the bottom layer of thatch is as old as the structure itself.

Thatching Today

What role does a thatcher play in America today? For the village thatcher of old, a project 20 miles from home was considered too far away. Today, I travel up and down the east coast of the United States, from Maine to Florida. In addition to the appeal of thatch for thatch's sake, thatch is gaining popularity because of a current trend in the United States toward using more natural building materials.

In the course of a year, I will roof outbuildings on an estate, roof a building designed to act as an Irish "pub," repair existing structures on homes and barns that set the stage for historic reenactments, and take on smaller projects in between. As thatch continues to regain popularity for both its visual appeal and its use of natural material, the need for skilled thatchers will only increase.

My Background and Work

While it may sound a little fanciful or pretentious, the title of Master Thatcher is a prestigious one that requires years of schooling, apprenticeship, and commitment to achieve. It also serves as a reminder of the serious approach and artistic skill that goes into a learned trade. Thatching and trades such as woodworking, glassmaking, and blacksmithing have all earned positions as practices that require a lifetime of dedication.

Having decided (at the tender age of 7) that I wanted to be a thatcher, my future was planned accordingly. When I was 16, my school counselor gave me the names of 50 thatchers working in England, and I began looking for an apprenticeship. In England, a 5-year apprenticeship, in addition to college courses, is mandatory before a thatcher can work alone. During my apprenticeship, I won my first award — for best journeyman/apprentice — for work on a roof in Essex. A few years later, I joined the prestigious East Anglia Master Thatchers Association. A year later, at 25, I was the youngest thatcher ever to receive the Ely Challenge Cup, for my work on a Tudor cottage.

After spending several years working in England, I was asked to take on a project in the United States. I'd always wanted to bring thatch back to the United States, and I accepted right away. The U.S. market for thatch has been steady, and I've been here ever since.

Master Thatcher Colin McGhee decided that he wanted to be a thatcher at age seven, when his father took him to see Robert Burns's cottage in Ayreshire, Scotland.

Making Thatched-Roof Birdhouses

The birdhouses in this book received significant special attention and research before I began making them. It was only after contacting various environmental and birding organizations, and building several trial structures, that I was able to come up with the designs I now market. All of the projects are designed to be practical and appropriate houses for a

This thatch is shaped to form what is called an eyebrow window.

particular bird (or bat or butterfly), as well as aesthetically pleasing and attractive additions to your yard.

These projects are simplified versions of thatching — they do not require any specialized tools or complicated techniques. Working on these small projects will give you a good introduction to the craft of working with thatch and creating a thatched roof. If you are interested in learning more about the complex thatching techniques used on larger structures, see the resources list provided at the back of this book.

▶ signing a roof

The ridge, or peak, of a thatched roof requires special attention because it is most vulnerable to leaks and damage. It often gets an additional cap of long straw to increase its thickness and create a slightly steeper angle.

There is a myth that each thatcher has a unique design he or she incorporates into the roof in order to "sign" a finished project and advertise for future work. In fact, ridge styles do vary across regions. Roofs in the eastern counties of England are angled or square, for example. Travel to the southwestern counties, and you'll see more gently sloping ridges and rooftops. The variety has more to do with the homeowner than the thatcher, however. The thatcher asks the customers what type of ridge design they want and offers suggestions as to what patterns best go with the roof.

Attracting Birds to Your Houses

The only way to be sure that the birds you want will use your house is to think like them. Success is never guaranteed, but these pointers will improve your chances.

▶ Keep a bird's normal nesting habitat in mind when picking a place to hang a birdhouse. If you live in a busy subdivision, you probably won't have a lot of luck with bluebirds; chickadees and nuthatches, however, may flock to your yard.

▶ Some birds are pickier than others about where they live, so be aware of their preferences. A list of each bird's preferences is included in each chapter.

▶ Although birds are not exceptionally tidy, they are more likely to nest in a place that doesn't have a lot of leftover nesting material in it. Removing an old nest also helps prevent parasites from making a home in the box and passing diseases on to the birds.

Each bird species makes a distinctive type of nest, using a variety of readily available and transportable materials.

Nesting sites are as varied as the birds themselves and reflect a species' preference for the ground or the treetops.

▶ Pay attention to the birds that live in your birdhouse. Some birds raise more than one brood a year, so make room for them by cleaning out a house once the previous nestlings leave. Each project in this book is designed for a specific species and describes the nesting habits of that species. Check to see when, and if, a particular family will be moving on.

▶ Birds are less likely to nest somewhere full of chemicals and fumes. Birdhouses offer limited ventilation, so avoid using any cleaning products or other chemicals inside the birdhouse. Because birds such as titmice seem to prefer natural cavities rather than birdhouses, keep the house as natural as possible in order to encourage them.

▶ Nothing draws a crowd like a free meal, so make your yard a little more bird-friendly by inviting birds to come and dine. While some species, such as bluebirds, prefer to stick mainly to insects, chickadees and nuthatches are fond of feeders.

▶ Perhaps the single most important thing you can offer birds is water. A bird bath or pond in a shady area is sure to be a hit!

▶ When using seed mixes, avoid rye, rice, oats, or other grains, as they attract starlings and blackbirds.

▶ Most birds that live in nesting boxes eat insects, although nuthatches and chickadees enjoy seed mixes and sunflower seeds throughout the year. Suet and seeds are especially helpful in the winter, when insects are harder to come by.

▶ Blue jays, cardinals, and other common backyard visitors always enjoy black-oil sunflower seeds or seed mixes. Squirrels, however, also have a preference for sunflower seeds or mixes that include them.

▶ Songbirds, such as goldfinches and siskins, don't care for nesting boxes but are still wonderful to listen to and watch. To attract these birds, offer them thistle seed in tube feeders. To deter housefinches from scaring off the other songbirds, look for an upside-down feeder. The birds must stand upside down in order to eat from one of these feeders, and this keeps housefinches away.

A pond or wetland is a good location at which to mount a box for swamp-loving birds.

▶ an inviting menu

TYPE OF BIRD	PREFERRED FOOD
Bluebirds, robins	Suet, bread, raisins, currants
Cardinals	Sunflower seeds, safflower seeds, fruit, bread
Chickadees, nuthatches, and tufted titmice	Suet, sunflower seeds, safflower seeds, peanuts
Goldfinches, siskins	Thistle seeds, sunflower seeds, safflower seeds
Orioles	Fruit pieces (especially juicy fruit, such as oranges), suet
Wrens	Suet, peanut butter, apples
Waxwings	Berries, currants, raisins, apples

An array of apples, oranges, raisins, peanuts, and sunflower seeds is sure to attract a variety of birds to your feeder.

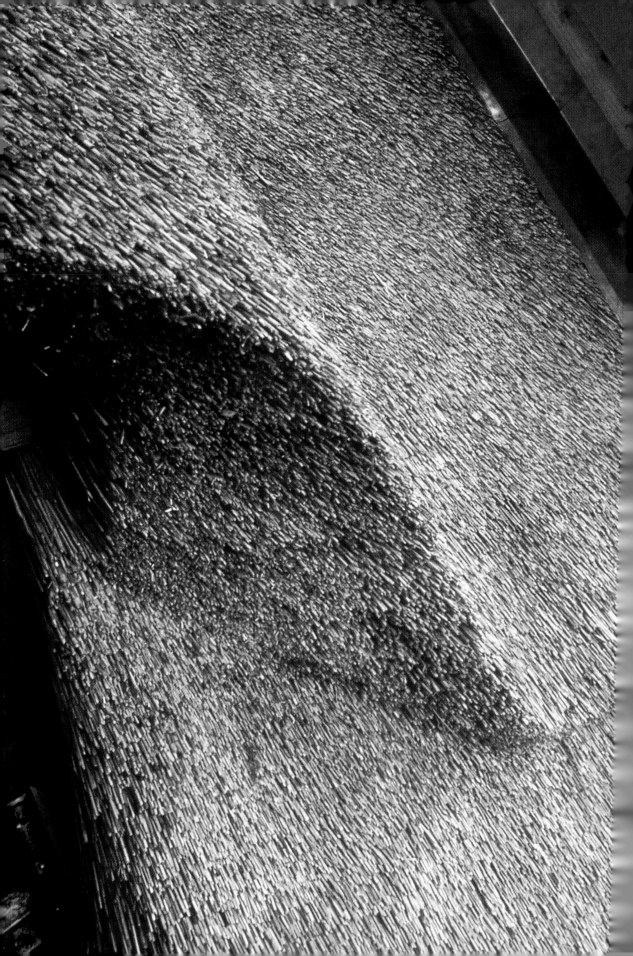

getting started: materials, equipment & techniques

If I ever become a rich man,
Or if ever I grow to be old,
I will build a house with deep thatch
To shelter me from the cold . . .

Hilaire Belloc

Thatching a house and thatching a birdhouse are slightly different propositions. A full-scale thatched roof is secured by iron hooks. If you visit a thatched-roof building today, you may be able to look up and see such hooks hanging down inside the building. The weight of the hooks and the other layers of thatch keep the material very firmly on top of the building. There are no adhesives in a real thatched roof, making them that much more impressive.

Thatch can be trimmed and shaped to create beautiful curvatures and lines in a roof.

Thatching Simplified

When it comes to anchoring the thatch to the roof for these projects, iron hooks are out of the question. Instead, I've added what I call anchors — strips of wood on top of the thatch that are screwed into a wooden base underneath. Wood and screws hold the thatch in place — no adhesives here, either.

How Much Thatch?

While a regular thatched roof uses "standard" bundles of thatch that measure 24 to 27 inches around, the projects in this book require smaller bundles, which I gauge by eye. The thatch on a birdhouse roof should form a layer 1 to 1½ inches thick (which will be slightly thinner after the anchors are screwed down). The anchors cannot hold more thatch than this.

Anchoring the Thatch

In order to work quickly and smoothly, you should have the anchors ready before you start thatching. Drill a ⅛-inch–diameter screw hole in each end of the wooden anchors, and start the screws. When you attach the anchors, drive the screws slowly so that no one screw or piece of wood puts all the pressure on the thatch. No matter how careful you are, however, the thatch is almost certain to loosen one of its anchors while you are working on it. Don't worry. Just move the anchor about ½ inch, and try again.

Tools You'll Need

Because most commercial thatching occurs overseas, it is difficult for Americans to locate the tools of the trade. Fortunately, these projects do not require that you write to England to obtain the shears and billhooks essential to making a real thatched roof. One tool that is necessary, however, is a pair of multipurpose scissors. These will come in handy for cutting thatch to the correct length and for clipping a decorative edge on the finished product.

Thatching can be done easily with the simple tools of a basic workshop.

The Basic Equipment

When I build these houses, I use all the tools in my shop, including a table saw and nail gun. You don't need a fully equipped shop, however. Most of the work can be done with simpler tools, and all the directions in this book are written for a less experienced craftsperson with a very basic workshop. Your tool kit should include the following, pictured above.

- Multipurpose scissors (**A**)
- Circular saw (**B**)
- Speed square for guiding the saw (**C**)
- Jigsaw (not shown)
- Electric drill with screwdriver bits and "woodeater" drill bits (**D**)
- Hammer (**E**)
- Handsaw or pruning saw (**F**)
- Tape measure (**G**)
- Sharp utility knife (**H**)
- Wedges (**I**)
- Screwdriver (**J**)
- Mallet (**K**)
- Hand pruning clippers (**L**)

getting started: materials, equipment & techniques

Important Safety Tips

Treat the circular saw with healthy respect. If you're not familiar with this tool, find someone who can teach you the ins and outs of working with it. Then buy some cheap pine and a good pair of sawhorses, and make practice cuts until you feel comfortable with the saw. The speed square is a standard carpenter's tool that helps you make straight cuts. It's a heavy-duty metal triangle with a lip that hooks over the board you're cutting. Guide the saw against the face of the square — one face results in a 90-degree cut; the other results in a 45-degree angle.

When you're working with the knife, always put whatever you're cutting on a firm, solid surface. (If nothing else, screw a hollow-core door to the sawhorses.) Attach small pieces to the work surface. Cut away from your body, and make sure your fingers aren't in the path of the blade.

Wear safety glasses when running the saw, and even when running the drill. It's not uncommon for a screwdriver bit to slip off the wood and swoop toward your eye.

Woodworking machines are loud, so protect your hearing with some inexpensive foam earplugs, available at most hardware stores, or safety ear muffs.

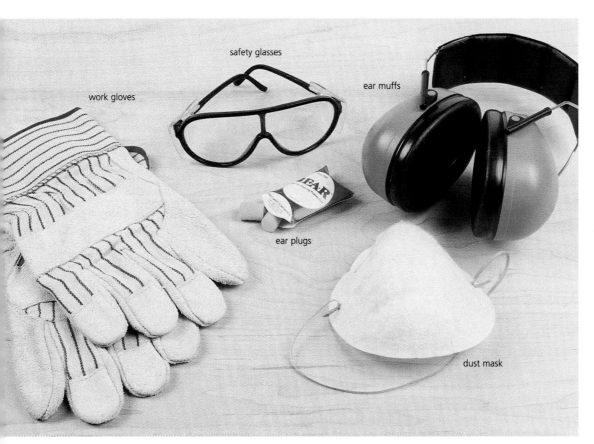

To work safely, be sure to use hand, eye, ear, and face protection.

▶ thatch-speak

Just to give you an idea of some of the terminology used in thatching, here are some definitions.

- **Bed:** A prepared/threshed heap of long straw, sedge, or rye.
- **Bottle:** A bundle (yealm) of straw tied at the small end and used to set eaves and gables.
- **Brow course:** First course of reed placed once the eave is set in; determines the pitch of the roof.
- **Bunch/bundle:** A unit of water reed approximately 24 to 27 inches in circumference.
- **Butt:** The lower end of a bundle of straw or reed.
- **Course:** A horizontal layer of reed or straw thatch.
- **Cross rods:** Hazel rods that are split and used for ornamentation between liggers.
- **Gable flue:** The finished edge of the thatch that hangs over the gabled end.
- **Gable top:** A yealm of ridging material without a pronounced taper at either end; forms the topmost part of the ridge.
- **Liggers:** Rods of split hazel used on the outside surfaces of ridges and, in the case of long straw, on eaves and gables.
- **Long straw:** Threshed wheat straw prepared by hand.
- **Pinnacle:** A raised end of a ridge or gable.
- **Ridge:** Capping on topmost part of roof.
- **Sweep:** Creating a valley.
- **Yealm:** A prepared drawn layer of long straw or sedge, 14 to 18 inches wide and 4 inches thick.

Material for Thatching

Cultures around the world have practiced thatching at some point, using whatever material was readily available — from straw roofs on Japanese temples to roofs of palm fronds on African houses. For the projects in this book, the long straw available at most craft stores is perfect. Make sure that the straw is unbroken, straight, and long enough for the particular project (most require 8- to 15-inch lengths, but the longest length called for is 20–40 inches).

Preparing the Thatch

Before cutting the thatch to length or placing it on the roof, hold the bundle loosely and tap the thicker end on a work surface. Pick out any pieces that are too long or too short, so that you have even lengths. This will save you time and effort later.

Straw tends to break, but you can make it more pliable and easier to work with by dipping it in water. It doesn't need to be sopping wet, just moist enough so that it isn't brittle. As you work, continue to moisten it. It will shrink as it dries, so you may have to retighten the anchor after 12 to 24 hours.

Cutting Your Own Thatch

Once you feel more confident about your thatching skills, try gathering and preparing your own thatch. While Norfolk reed is the choice of thatchers along the English coast, other types of hollow reed or grasses also work well. Take the time to notice areas near you that have plants that interest you. When the time comes to harvest the plant, make sure you have permission from the landowner before cutting anything down. Many wild areas are protected and need grasses or reeds left in place to prevent erosion, so it is important to get permission before you cut.

What kinds of plants work? The most important thing is that the material be hollow so that it will channel rain down the roof. The hollow stems inside cattails are a good, commonly available source. Broom sage and other tall, stiff grasses also work well.

Selecting and Drying

You want your materials to be thoroughly dried before you thatch. Winter, when you can find brown, dried plants still standing in marshes or meadows, is the best time to harvest thatch. Choose older, mature plants that have the thickness and strength you will want when securing the stems to a roof. Try to harvest a little more than you need, in case you make a mistake or run short.

It is possible to use green plants if you let them dry out before use. Look for thick, heavy stems closer to the bottom of the plant. Material shrinks slightly as it dries, so cut your pieces a little long. Lay out the material you're using in a warm, dry place. Spread the grass or reed in as thin a layer as possible, and turn it every day to prevent the bottom layer from mildewing. For better results, lay the thatch on a raised screen that is supported by a wooden frame. Keep the stems straight and parallel — a pile of bent, broken stems won't work. Most important, be patient during this process. The more attention you pay to the thatching material, the better the chances it will stand up to the process and result in an attractive, well-crafted roof.

MASTER THATCHER tips

Exercise creative freedom when trimming the thatch edges: Cut the thatch in a simple curved- or scalloped-edge pattern along the bottom of the roof. Use multi-purpose scissors (Fiskars makes a version with cushion-grip handles) and remember:

- Cut the top layer of each section in small amounts, about half the thatch's thickness. After cutting a small section from the top, cut the layer underneath in the same lines.
- Cut a piece from one side, then the other, until you reach the middle. This keeps the pattern symmetrical.

Building Materials

It's important to consider what kind of wood you are going to use to build each of your birdhouses. I recommend avoiding treated woods since most of the preservatives used in these woods are harmful to birds. As you will note in the materials list for each box, I prefer to use cedar, poplar, and plywood. For several designs, I harvest much of my own wood, and I create accent pieces from tree limbs found near my house. I encourage you to use what is readily available around your yard.

You'll need the usual hammer, nails, and screwdriver for building your thatch projects, but consider carefully the kind of wood you'll use — untreated pine, cedar, poplar, and plywood are best, as they do not have preservatives that are dangerous to birds.

learning to thatch

In Architecture as in all other Operative Arts, the end must direct the Operation. The end is to build well. Well building hath three Conditions. Commodity, Firmness, and Delight.

Sir Henry Wotton

If you'd like to get a little bit of experience thatching before you start building birdhouses, you can try thatching something that someone else has already built. A commercially available bird feeder is a great place to start.

The advantages of beginning with an existing structure are several. First, you can focus entirely on thatching and become proficient at it without the work of first building the base structure. Also, you will finish your thatching project quickly and will be energized to start another, probably more complicated, piece. Last, perhaps you don't have the skills or the interest to make the base structure; in that case, you will still be able to produce unique thatched pieces that will be beautiful and satisfying!

In this chapter, I explain thoroughly, from beginning to end, the process of thatching a roof. Later, in each of the projects in this book, I'll explain the unique processes for each piece, but you will be referred back to this chapter for complete thatching directions. Let's get started!

A store-bought bird feeder can be transformed into a one-of-a-kind piece with the addition of a thatched roof.

Selecting a Feeder

Before we get under way, let's talk about the kind of feeder best suited for thatching. Incidentally, I recommend thatching a bird feeder instead of a birdhouse, because feeders are usually well-built and work well. Commercially available birdhouses, on the other hand, are usually one-size-fits-all. Given the differences between, for example, an owl and a chickadee, I like to custom-build houses to fit the needs of individual species.

> ▶ **for decoration only**
>
> If you want to make a decorative thatched house for display indoors as part of your home decor, you can use any type of commercially available birdhouse. Once it's thatched, you might want to add decorative painting or other decorative elements.

Most garden centers have beautiful wooden bird feeders that can serve as the base for your first thatching project.

It goes without saying that the feeder should be made of wood. Thatch would not only look silly on a plastic feeder, but would be hard to attach to it. Hooks — or other hanging hardware that sticks from the roof — can be tricky to thatch around, so if you want the easiest project, I recommend buying something you can mount on a pole. Get a feeder that has a roof with two faces — like the roofs on most houses — so that you can practice aligning the straw along the ridge.

Look for the steepest roof you can find. A traditional thatched roof has a 45-degree slope so that rain will run off quickly. The slope isn't quite as important here — you'll be applying thatch over the roof that's already on the feeder, and presumably, it's already watertight. The 45-degree pitch is as important to the appearance of a thatched house as it is to its function, however, and you do want the roof to look right. A thatched roof with a flat pitch is going to look as out of place as a modern hot-tar roof on a castle. You can get away with it, but people are likely to notice that something isn't quite right.

One way to make the roof edges for the feeder is to split a branch in half lengthwise.

Thatching a Roof

There are three basic phases in thatching this roof: applying the edges, applying the thatch, and anchoring the thatch.

Preparing the Edge Strips

All the projects in this book have strips of wood nailed to the edge of the roof. The strips extend above the roof so that you can nest the straw against them when you're thatching.

To make the edges for the feeder, you will need four pieces of wood, each about ½- to ¾-inch thick, about 1½ inches wide, and 1 inch longer than the side of the roof. You can either split a branch in half lengthwise or buy prepared wood at a hobby shop.

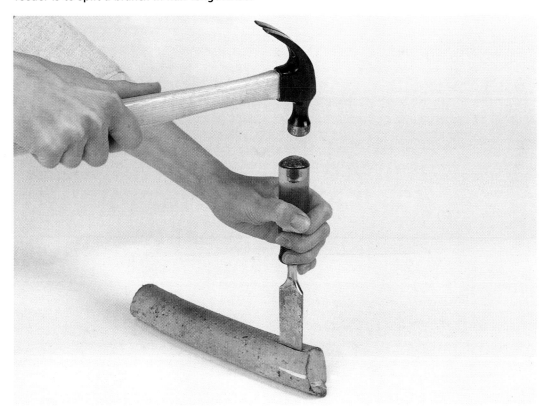

learning to thatch

Preparing the Thatch

The directions for each project measure thatch by the handful — a bundle about 3 inches in diameter. To measure out a handful, grasp a bunch of thatch around the middle. Hold it just tightly enough to keep it from spilling to the floor, but not so tightly that the tips of your thumb and middle finger meet.

For this project — or any other, for that matter — you'll want one handful for every 3 inches of roof width. If a roof is 9 inches wide and has only one side (like some sheds), you'll need three handfuls. If the roof is 9 inches wide and has two sides (like most houses), you'll need six large handfuls. Thatch should hang over the edge of the roof by 3 to 5 inches. It is better to let it hang long and trim it back as needed.

Before cutting thatch to a desired length and placing it on the roof, hold the bundle loosely and tap the thicker end of it on a work surface. This gives you even lengths of thatch and saves time and effort later in the process.

Elsewhere I suggest that you keep your eyes open for suitable thatch material that grows in your area. Until you've found something, you'll probably be working with straw, a traditional thatch material.

Because straw tends to break, make it more pliable by dipping it in a bucket of water. It doesn't need to be sopping wet, just moist enough to keep from being brittle. As you work with the straw, continue to moisten it if it dries too much. Keep in mind that it will shrink as it dries, so after 12 to 24 hours, you may have to retighten the anchors that are holding it in place.

Preparing the Anchors

The wooden anchors you use to screw the thatch to the roof should be the width of the roof and ½- to ¾-inch thick. You will need two anchors for each face of the roof on this feeder. Space them horizontally across the thatch, roughly 2 inches apart.

Be sure to prepare the anchors in advance. I make them by splitting branches or dowels lengthwise; the flat side goes against the thatch. The directions throughout this book recommend splitting the stock with a 3-pound sledgehammer and the kind of wedge you use to split firewood. (A 3-pound sledge has a handle about the size of that on a regular hammer and works well for splitting small pieces.)

Splitting the pieces can be a bit of a hit-or-miss proposition, depending on the grain of the wood. Straight-grained pieces of oak, ash, hickory, or Chinese chestnut will split cleanly and evenly. Tighter-grained woods, or woods with knots or wild grain, may not split as nicely. While a split that's hard to control may be a nuisance, a piece that's a bit uneven will still do its job and doesn't look the least bit out of place on a rustic birdhouse. If you're fussy, you can cut the pieces on a band saw or a pruning saw. The first option obviously requires a band saw; the second requires some patience and muscle.

Predrill holes for the screws. The holes should be slightly larger than the diameter of the screws. Drill one in the middle of the anchor and one about an inch from each end. Put the screws in the holes in advance, to minimize fumbling when it comes time to secure the straw.

Thatching the Feeder

With your feeder, thatch, and anchors prepared, you are finally ready to begin thatching! Once you've mastered this basic technique, you'll be able to apply it to all the designs in this book — and to your own designs, as well, if you feel creatively inclined.

Applying the Roof Trim

Each trim strip should be about 1½ inches wide. The length depends on the size of the roof. Cut four pieces, each about an inch longer than the edge of the roof.

Each end of the trim is cut at an angle, with the angled ends meeting at the top of the roof. Since the angle varies depending on your bird feeder, it's best to nail the edging in place first, and then cut the angle.

Begin by aligning an end of the edging with the lower end of the roof. Position so it forms a 1-inch lip, as shown. Nail it in place with 1-inch brads.

Nail a second edge piece in place, letting it overlap the first at the ridge. Cut through both with a saw to form a neat seam at the ridge.

Thatching One Face of the Roof

Prop the bird feeder against a piece of wood or other object so that one face of the roof is parallel to the surface of the table you're working on.

Place small handfuls of thatch on one face of the roof, patting it in place as you work and allowing the bottom edge to hang over the lower edge of the roof trim by 1 to 2 inches.

Work with small handfuls of thatch on one side of the roof and pat them into place as you go.

It's best to nail the edge pieces in place first, then cut them at the ridge, to ensure a neat seam.

Anchoring the Thatch

When the thatch is about 1 inch thick, place one of the wooden anchors on the top of the thatch and screw it lightly in place, but not so tightly that you can't move the thatch back and forth.

Evening Out the Thatch

Using a hammer or piece of wood, lightly tap the edges of the thatch along the top of the roof to line it up with the wooden roof.

Tap the thatch at the top of the roof lightly to create an even peak.

When the thatch is placed, screw the anchors down lightly — give yourself room to move the thatch back and forth if necessary.

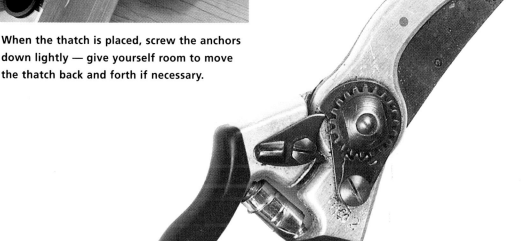

Thatching the Second Side of the Roof

Reposition the birdhouse, propping it so that the second side of the roof is parallel with the surface you're working on. Apply thatch as you did to the first side, placing small handfuls, and pat it into place until the thatch is about 1 inch thick. When it is, apply one of the wooden anchors, and screw it lightly.

Aligning the Ends of the Thatch

Using a small block of wood, gently begin tapping along the bottom edge of one side of the roof, then along the other side, gradually evening the bottom edge of the roof and bringing the top edges together.

Lightly tap the top of the roof to create an even peak along the ridge of the roof. Once the thatch is even, tighten the screws in the anchors, and screw the remaining anchors in place.

Trimming the Top of the Roof

Once you have evened up the thatch at the top to create a peak, you can use multipurpose scissors to snip off any errant ends.

Trimming the Edges of the Roof

You can cut a pattern into the roof, the same way thatchers trim full-sized roofs. Multipurpose scissors are perfect for making the cuts.

Start trimming by cutting small amounts at a time from the top layers. Usually a ½-inch layer works well, but adjust the cut to an easily manageable amount.

To keep the pattern symmetrical, cut a section of the pattern on one end of the roof, and then cut a matching section on the other side. Continue cutting and matching until you've reached the middle of the roof.

By using multipurpose scissors, you can create a professional-looking pattern in the bottom edge of your thatched roof.

Fill your finished feeder with birdseed and mount it near your house so you can observe the birds that gather.

thatched bird feeder

At the corner of Wood Street, when daylight appears,
Hangs a thrush that sings loud, it has sung for three years:
Poor Susan has passed by the spot, and has heard
In the silence of morning the song of the bird.
William Wordsworth

This attractive feeder is just the thing for attracting birds to your yard. In addition to having a thatched roof and rustic appearance, this feeder is large enough to support a variety of birds at once. The open tray is easy to fill and to clean out, as needed. See page 11 for a guide to selecting the feed that will attract the birds in your area.

Hang this feeder near a few birdhouses, and you'll be impressed with the variety of birds you can attract. Be sure the feeder is in a spot you can easily observe from a comfortable place in your house so you can fully enjoy the activity.

The thatched-roof bird feeder is one of the easiest projects to build and is a great way to learn the roof-thatching technique.

▶ a word about squirrels

It seems that the squirrel never tires of discovering creative ways to get into a feeder. Backyard birders have tried to come up with homespun techniques to keep these furry tricksters out of their feeders, only to be foiled time and time again. While the final word from many people is "Learn to live with them," there are a few steps you can take to make things more difficult for squirrels.

Assume the squirrel's point of view. The squirrel, which hunts and pokes around all day for food, suddenly sees a tray full of seeds within easy reach. He doesn't know that you don't want him there. He just wants all that limitless, available food. And he has three main ways of reaching it.

1. Climbing up. Many people attach a plastic dish or a couple of aluminum pans to the feeder pole to prevent squirrels from climbing up. If you expect this to work, don't do the job too well: If the flashing is sturdy enough to support a squirrel's weight, he can climb over it. Your best bet is a strong, heavy flashing that's hard for the squirrel to grip but that is mounted loosely, so that it will tip if a squirrel does get on top of it.

Another solution, if you're at the end of your tether, is to grease the pole. Or, more simply, get rid of the pole altogether. Which brings us to the squirrel's second method.

2. Climbing down. A feeder hanging from or near a tree or other structure is just begging squirrels to test their skill by climbing down to reach the food. Flashing works well here, too, provided that it is hard enough to prevent squirrels from getting a claw-hold and that it wobbles enough to dump them unceremoniously on their heads.

3. Jumping. This is the trick most backyard birders overlook. We move the feeder away from a tree and hang it from a pole the squirrel can't climb, and the next thing we know, the squirrel's in the feeder having a feast. Squirrels are amazing acrobats. A feeder within jumping distance is an easy mark.

My father came up with two techniques that have worked well. The first is to stretch a heavy piece of fishing line (not rope — squirrels can climb across it) between two widely spaced trees. Hang the feeder from this line so that it is suspended in space, out of jumping range. The other trick is to provide a corn feeder for squirrels on the opposite side of the yard. The squirrels seem to prefer the corn, and my Dad hasn't had any squirrel woes since trying it.

building the feeder

Difficulty: Easy

This is one of the quickest projects to build since it does not require constructing closed sides, a closed front, or a closed back. The feeding tray and rims are made from cedar or pine; you can use plywood for the roof panels. The remaining pieces can be cut from small logs found in your yard or from large wooden dowels.

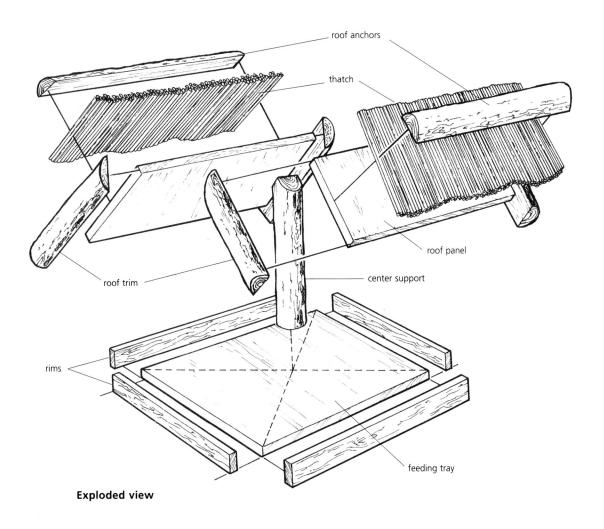

Exploded view

▶ what you'll need

▶ Parts

- Feeding tray: 2 x 12* of cedar or pine, 11¼" long
- 2 rims: 1 x 3† of cedar or pine, 11¼" long
- 2 rims: 1 x 3† of cedar or pine, 12¾" long
- 1 roof panel: ½" x 4½" x 13"
- 1 roof panel: ½" x 5" x 13"
- Center support: 2"-diameter log, 10" long
- Roof trim: 1½–2"-diameter log, 7" long
- Roof anchors: make from 1½–2"-diameter log, 13" long

*The true size of a 2 x 12 is 1¾" x 11¼".
†The true size of a 1 x 3 is ¾" x 2½".

▶ Materials

- 4d (4-penny) nails
- Exterior wood glue
- Enamel spray paint (dark color)
- 8 large handfuls of thatch, 8" long
- 2" galvanized deck screws
- Brads or other small nails
- Large staple or screwable eye to attach chain if hanging (optional)
- Hanging chain: 10"–12" length for hanging feeder (optional)
- 2" square post, at least 6' long for post mounting (optional)

▶ Tools

- Circular saw
- Hammer
- Pruning saw or handsaw
- Drill with screwdriver bit and ⅛" and ⅜" drill bits
- Multipurpose scissors
- Firewood wedge
- 3-lb. sledgehammer

Assembling the Feeder

1. Nail the rim pieces to the sides of the tray with 4d nails. ▼

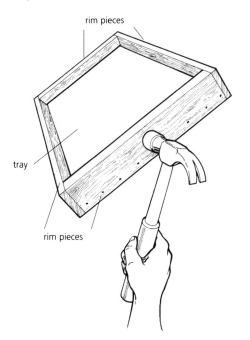

2. Assemble the roof by putting the end of the 5-inch piece over the edge of the 4½-inch piece to form a 90-degree angle. Glue and nail the pieces together with 4d nails. Spray-paint the underside of the roof with a dark color to protect the wood. The paint also helps disguise the plywood and keeps the focus on the thatch rather than on what is underneath it.

3. The center support is a log with a point on one end that nests inside the roof. On the 2-inch-diameter log, mark the approximate center on the end grain. Cut a 45-degree angle from the mark to the edge of the log with a pruning saw or handsaw, gauging the angle by eye. Repeat, cutting from the center to the opposite side. ▼

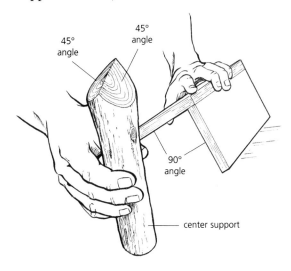

MASTER THATCHER tips

To make the roof panels, start with a piece of plywood 13 inches wide and 9½ inches long. To cut, guide the edge of a circular saw against a straight piece of wood temporarily screwed to the plywood. Crosscut the strip into two pieces, one 4½ inches long, the other 5 inches long, guiding the cut with a speed square.

Attaching the Center Support to Roof and Tray

1. Glue and center the pointed end of the support to the center of the underside of the roof. Nail it in place with 4d nails. ▼

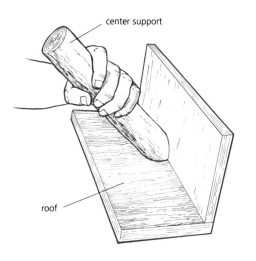

2. Find the center of the food tray by drawing lines diagonally from corner to corner on the underside. Drill a ⅛-inch hole through the tray where the lines cross. ▼

Drill about 10 drainage holes of ⅜-inch diameter into the bottom of the food tray, keeping them at least an inch away from the center hole.

3. Drive a 2-inch galvanized deck screw through the center hole and into the center support.

Drill a couple more holes through the bottom of the tray, and just into the log. Drive the deck screws through the hole and into the support to attach it securely.

Drill two holes through each side of the roof and just into the log. Drive two 2-inch screws through the holes, securing the roof firmly to the log.

Attaching the Roof Trim and Thatch

1. Split a 1½"- to 2"-diameter log in half, using a firewood wedge and a 3-pound sledge. Cut a 45-degree angle on one end of each, using a handsaw or pruning saw. Nail the trim to the roof edges, as shown. ▼

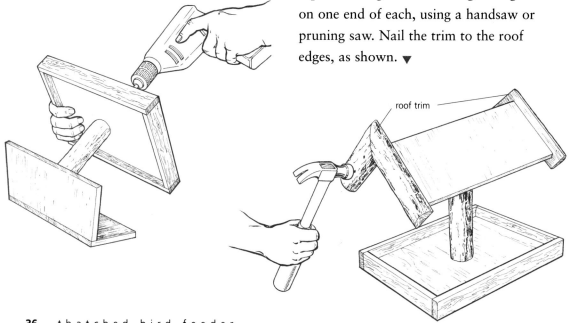

36 thatched bird feeder

2. Place small handfuls of thatch on one face of the roof, patting it into place as you work, allowing the thatch to hang below the lower edge of the roof trim by 1 to 2 inches. When the thatch is about 1 inch thick, place one of the wooden roof anchors on the top of the thatch and screw it in place, but not so tightly that you can't move the thatch back and forth.

3. Thatch the second side the same way. Align the bottom edges of the thatch by tapping with a wooden block, and then tap along the top of the roof to create an even peak. Tighten the screws holding the anchors, and trim the thatch with multi-purpose scissors. (For complete directions, see Thatching a Roof on page 23.)

4. Spray the roof with a clear enamel spray gloss or waterproof lacquer. This gives visual appeal and will also help prevent the thatch from shifting over time.

Hanging the Bird Feeder

To hang: Attach a screwable eye to the center of the roof ridge (pushing thatch aside slightly) and attach chain to eye.

To mount: Dig a hole, and sink a 2-inch-square post 6 inches into the ground for every foot it will be above the surface. Drill four pilot holes in the bottom of the tray, and screw the feeder to the top of the post with 2-inch galvanized deck screws.

Using hand-split logs for the roof trim and anchors adds rustic charm to your feeder and gives you some leeway if parts don't fit together exactly.

chickadee box

*Ah, sad and strange as in dark summer dawns
The earliest pipe of half-awaken'd birds
To dying ears, when unto dying eyes
The casement slowly grows a glimmering square . . .*
Alfred, Lord Tennyson

Almost anyone who's ever put birdseed out in the yard is familiar with the black-capped chickadee. One of the most consistent visitors to backyard bird feeders, the chickadee is a miniature acrobat, capable of hanging upside down from branches to pick off every bit of food or diving through the air in pursuit of an insect.

Black-capped chickadees are flocking birds, which may explain their amiable, chipper nature. They gather at bird feeders as a small flock, and each bird seems to wait its turn for a nibble.

This box is designed to mimic a tree cavity, the favorite nesting spot of black-capped chickadees. It can be easily mounted on any tree in your yard.

Built of poplar, cedar, or yellow pine, this simple box can be left unfinished and allowed to weather naturally.

The Black-Capped Chickadee

Black-capped chickadees *(Poecile atricapillus)* are small birds, weighing a little under half an ounce. Despite their size, they make a lot of noise, and their familiar chick-a-dee-dee-dee call is one of the loudest in the backyard. Chickadees use their call as a way to keep the flock together, so it must be loud enough for other birds to hear.

In addition to their familiar black caps, chins, and throats, chickadees have gray backs, tails, and wings and white breasts. Male, female, and young birds all have similar plumage.

Chickadees range over a large part of the United States. They prefer mild to cold climates and can be seen as far north as Alaska, and south into the mountains of Virginia and Tennessee. Because they do not migrate to warmer climates as the seasons get colder, black-capped chickadees are one of the most common birds seen during the winter.

▶ the bird statistics

The Black-Capped Chickadee at a Glance

Length: Between 5 and 6 inches

Diet: Insects, spiders, caterpillars, snails, slugs, centipedes, seeds, berries, and suet

Habitat: Mixed deciduous and coniferous forests, both deep inside the forest and along its perimeter. Can be found in more open areas with scattered trees, such as old meadows or parks. Found in both rural and suburban areas.

Region: Northern two-thirds of the United States. Southern and western Canada, stretching as far north as Alaska.

Breeding season: Early April to early May, depending on climate

Nesting Requirements

Entrance hole: 1⅛ to 1½ inches in diameter. Hole should be about 6 inches above the box floor.

Height of box above ground: 4½ to 15 feet. Install the box so that it is accessible for cleaning.

Location of box: In clearings or along the edge of a forest or wooded lot. The box should face south so it receives sunlight at least half of the day.

Chickadee Diet

Living in colder climates means taking advantage of many different food sources. As the seasons change, chickadees have their choice of a variety of food sources. Insects make up a large portion of their diet, but chickadees will also forage for berries and seeds. An additional food, which may help keep the birds warm in chillier weather, is the fat from dead animals. Suet is a popular alternative in many backyard feeders.

Like other animals that tough it out during the winter, chickadees adapt to food shortages by "caching." They hide berries and seeds in trees between cracks in the bark or under leaves, and they come back for them later. Studies on this behavior reveal that chickadees can remember the location of cached items for up to a month.

Mating and Nesting Habits

Chickadees are monogamous and stay together for life. As soon as the birds pair up, they claim a territory and defend it with amazing assertiveness. They will defend the area as long as both birds in the pair are alive, although they do choose different nesting sites within the territory for each new brood. If one of the pair dies during the winter, the survivor will choose a new mate in the fall, when the winter flock is taking shape. In the spring, the search for a nesting site begins again. The chickadee breeding season begins in early April in warmer locations and in early May farther north.

Nesting birds select a site in a tree cavity, a hole drilled by a woodpecker, or in a birdhouse, and then spend seven to ten days cleaning and preparing the site for a nest. The female builds the actual nest in about three to five days, al-though it sometimes takes her a couple of weeks. She makes it out of moss and lines it with soft, warm material, such as rabbit fur, hair, down, plant down, and spider webs.

The Family

The female chickadee lays one egg a day, usually in the morning, and covers the eggs with nest material whenever she leaves. A clutch usually has six to eight small, pale pink, nonglossy eggs that are covered with red, brownish, or purple specks. Once all the eggs are laid, the female remains with the clutch to incubate them — a period lasting about 12 days.

After the chicks hatch, the female remains with them for another few days while the male bird brings food. Within a few days, the female begins to leave the nest and help with the feedings. The young birds lose their downy coats and fledge at 16 days, although their parents continue to feed them for two to four weeks after they leave the nest. Chickadees have only one brood per season.

building a thatched chickadee box

Difficulty: Easy

Constructed of cedar or pine, this natural-looking box is designed to blend in with the surrounding trees. The angled front piece and triangular sides fit together neatly in this easy-to-assemble pattern.

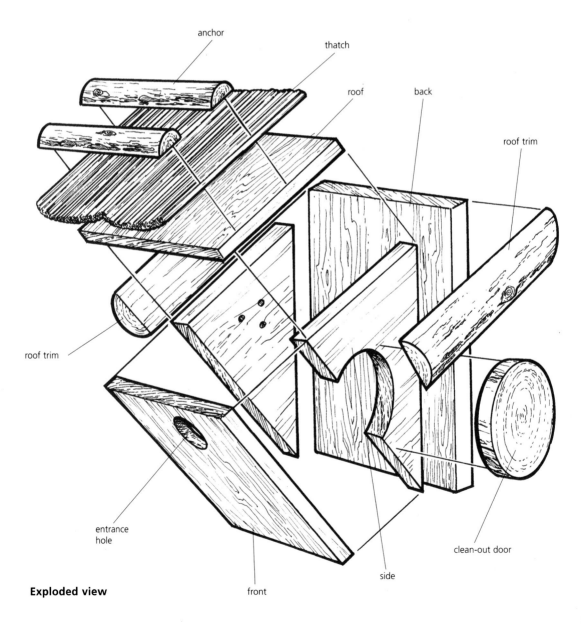

Exploded view

▶ what you'll need

▶ Parts
- Back: 1 x 8* of cedar or pine, 14" long
- Front: 1 x 8* of cedar or pine, 8" long
- 2 sides: make from 1 x 8* of cedar or pine, 30" long
- Roof: ¾" x 7¼" x 7¼" piece of plywood
- Roof trim: 2–3"-diameter log, 10" long
- Clean-out door: ½"-thick slice of wood, cut from a 3–4"-diameter log
- 4 roof anchors: make from 1½–2"-diameter logs or dowels, each 7¼" long

*The true size of a 1 x 8 is ¾" x 7¼".

▶ Materials
- Sandpaper
- Exterior wood glue
- 1" brads
- 4d (4-penny) nails
- One 1¾" galvanized screw and washer for clean-out door
- Four 2" galvanized screws
- 3–4 large handfuls of thatch, 12" long

▶ Tools
- Circular saw
- Speed square
- Compass
- Pencil
- Jigsaw
- Sandpaper
- Piece of ½" dowel
- Hammer
- Firewood wedge
- 3-lb. sledgehammer
- Pruning saw or handsaw
- Drill with screwdriver bit and ⅜", ½", and 1½" drill bits
- Multipurpose scissors

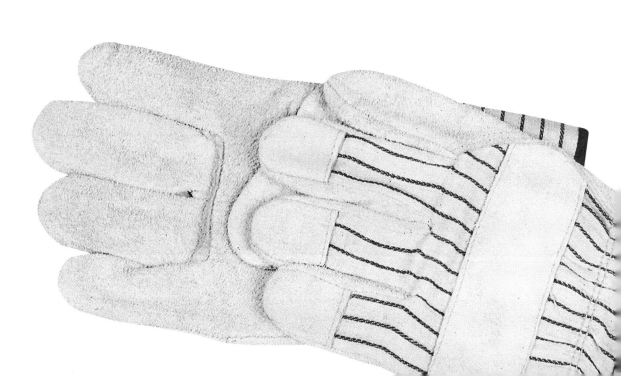

Cutting the Sides and Assembling the Chickadee Box

1. When cutting the triangular sides of the chickadee house, begin with a piece of 1 x 8 stock that's at least 30 inches long. Guide your circular saw along the diagonal of a speed square to cut a right triangle with two 7½-inch sides. Then guide the saw along the straight edge of the speed square to square off the stock. (The scrap left in between is too small to use as the second side.) Cut a second triangle in the same manner. ▼

2. With a compass and pencil, lay out a half circle, with a 1-inch-diameter radius, along one of the short sides of one of the triangles. This will be the hole to use for cleaning the birdhouse. Cut along the layout line with a jigsaw. Wrap sandpaper around a dowel, and sand the inside of the clean-out hole smooth. ▼

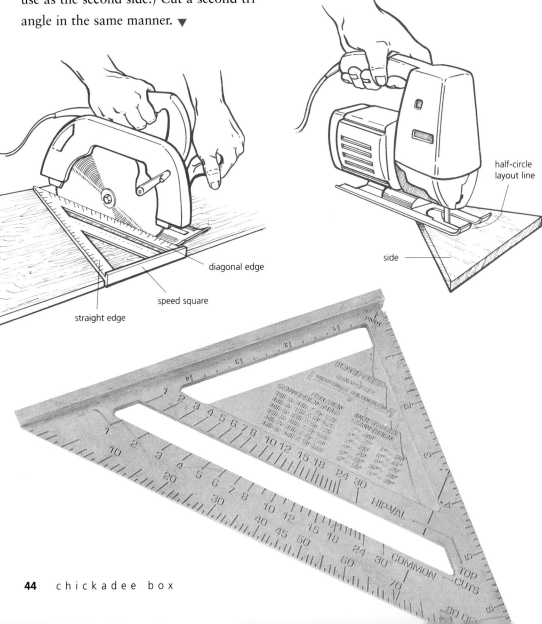

3. Draw a layout line across the back, making it 1 inch from the bottom edge (the shorter dimension). With the back piece lying flat, put the long edges of side pieces on top of the back, aligning the bottom edge of each piece along the layout line. Position the sides so the long edge of each is against the back and the cleaning hole faces the bottom edge. Glue and nail the sides in place.

Put the front in place, and glue and nail it to the sides (refer to exploded view, if necessary). ▼

Attaching the Roof and Trim

1. Position roof piece over the top of house. Glue and nail it into place. ▼

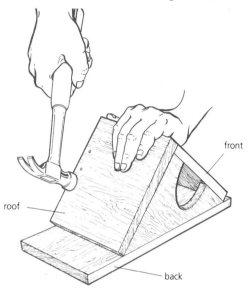

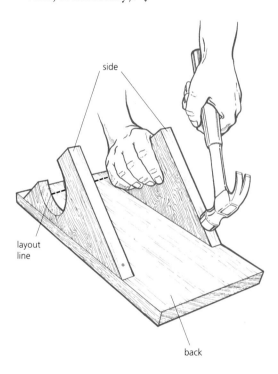

2. Split in half (lengthwise) a 10-inch-long log, using a firewood wedge and a 3-pound sledge hammer. With a pruning saw, cut a 45-degree angle across one end of each half. Nail both pieces of the log onto the birdhouse, lining up the top of each with the top of the back of the birdhouse. There should be a 1- to 1½-inch lip above the roof to support the thatch. ▼

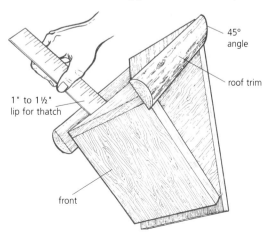

chickadee box 45

Drilling Holes and Attaching the Clean-Out Door

1. Drill a 1⅛-inch-diameter entrance hole in the center of the front, 1 inch down from the roof. Sand the inside of the hole with a bit of sandpaper wrapped around a piece of dowel.

Drill three ¼-inch-diameter ventilation holes in the solid side, laid out in a triangle. Leave at least 1 inch between the holes and the opening. ▶

Drill a ½-inch-diameter mounting hole in the back, near the top.

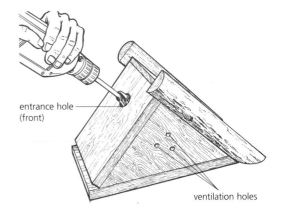
entrance hole (front)

ventilation holes

2. Place the clean-out door over the clean-out hole, and drill a ⅛-inch hole through it, near the top, into the side of the birdhouse. Screw it in place using a 1¾-inch screw and washer. Fasten it for a snug fit that still allows the door to swivel open for cleaning. ▼

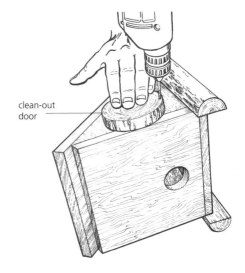
clean-out door

The clean-out door is mounted with a screw and washer so that it swivels open for easy cleaning.

Applying the Thatch

1. Place small handfuls of thatch on the face of the roof, patting it into place as you work and allowing an overhang of 1 to 2 inches beyond the roof trim edges. When the thatch is about an inch thick, place one of the wooden anchors on the top of it and screw it lightly in place, but not so tightly that you can't move the thatch back and forth.

2. Align the bottom edges of the thatch by tapping with a wooden block, and then tap along the top of the roof to create an even peak. Tighten the screws holding the first anchor, and firmly attach the remaining anchor. Trim the thatch with multipurpose scissors. (For complete directions, see Thatching a Roof on page 23.)

3. Spray the roof with a clear enamel spray gloss or waterproof lacquer. This is mostly for visual appeal but will also help prevent the thatch from shifting over time.

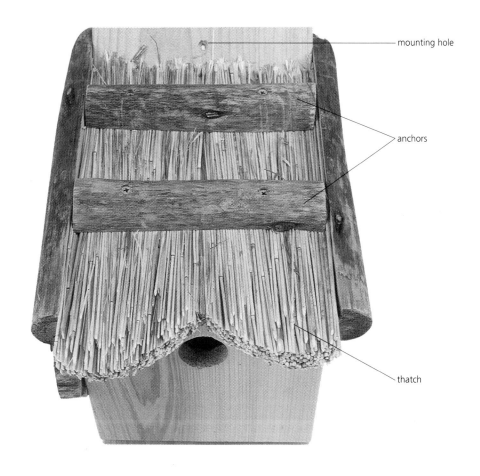

screech owl house

*Sweet the coming on
Of grateful evening mild, then silent night
With this her solemn bird and this fair moon,
And these the gems of Heav'n, her starry train.*
John Milton

There are two common varieties of screech owl in the United States, the eastern and the western. Both are fairly small (for owls, anyway) and are usually heard more than seen. One or the other can be found almost anywhere in the United States, particularly in lower elevations.

Since owls are nocturnal, you may hear them more often than you see them. However, by mounting a nesting box in your yard, you increase your chances of catching sight of one of these stately birds. You might see one perched on a branch at dusk quietly scanning for mice and other rodents on the ground.

This house requires that you work with some larger pieces of wood, but it is still relatively easy to assemble.

The Eastern Screech Owl

the bird statistics

The Eastern Screech Owl at a Glance

Length: From 6¼ to 10 inches. Males are normally 2 to 4 inches smaller than females.

Diet: Insects, small birds, rodents, frogs, toads, small reptiles, and fish

Habitat: Wooded areas below 4,875 feet in elevation. The eastern screech owl is found in a variety of habitats, including mixed forests, swamps, farms, orchards, suburban neighborhoods, and parks.

Region: Eastern two-thirds of the United States

Breeding season: Begins late January to early March, depending on climate

Nesting Requirements

Entrance hole: About 2¾ inches in diameter, about 9 inches above the floor

Height of box: 15 to 50 feet

Location of box: Choose a shaded area that doesn't have too many overhanging branches, so that the owls have a clear flight path to and from the box. If possible, hang the box on a tree that is wider than the box.

Despite its name, the eastern screech owl (*Otus asio*) is more often known for a variety of nonscreeching noises, from easily recognized hoots to barks and rasping sounds. Until recently, the eastern screech owl population was on the decline as more and more land was cleared. In response, however, the bird seems to have extended its range farther west. Attempts to reforest land and invite the screech owl's return have been helpful, but nesting boxes are still necessary, since young trees don't have many of the natural cavities that owls nest in.

Both female and male eastern screech owls have prominent ear tufts that they can raise or lower; the owl's head is round when the tufts are lowered. Screech owls may be red or brown. The red bird is more common in the South, while the brown variety is common in the North and in southern parts of Texas. Both types have bars and streaks across the underside and breast, and both have large, yellow eyes.

The eastern screech owl's range extends from the Atlantic Ocean to the edge of the Rocky Mountains and the east coast of Mexico. Despite the wide range of their habitat, these birds have been forced by human and land development to adapt to even more types of environments. While a mixed deciduous forest may provide more natural surroundings, the owls are gradually learning to be equally at home on a farm or in a suburban park. However, screech owls do prefer that the sub-canopy (the portion of forest below the branches) not be too dense, so that they can fly and seek out prey.

Eastern Screech Owl Diet

Like all owls, screech owls are predators, with a diet consisting of small mammals, reptiles, amphibians, and insects. Owls will also catch smaller birds and, if they are available, crayfish and other small fish. They're nocturnal hunters, and the bars on their feathers help camouflage them in the trees. Owls sit perfectly still, their large eyes scanning the surrounding area for food. When prey is in sight, the owl swoops down. A layer of soft feathers under its wings keeps the animal below from hearing its approach.

▶ screech-owl predators

Although the eastern screech owl is a predator, it faces many of the same risks other birds do when it comes to finding a safe nest. Fox squirrels, kestrels, and flickers will often take over a nest — the squirrels by eating the eggs, and the birds by simply pushing the eggs out of the nest. Like their eastern counterparts, western screech owls face predation from egg-eating rodents and other birds, as well as nest-site competition from the kestrel.

Screech-owl eggs and babies are also popular with rat snakes, opossums, and raccoons, all of which will climb into a nest and feast. Larger owls and even other screech owls will sometimes invade a nest to eat adult birds and their young.

One way to help protect screech owls from attacks is to provide guards on the tree or pole where you are mounting the nest. Predator guards usually take the form of a wide ring of wood or metal that prevents an animal from climbing up the pole and reaching the nesting box. Place the guard about halfway up the pole, making sure that there are no branches or other objects between the guard and the box that would still offer access to the nest.

Mating and Nesting Habits

Eastern screech owls are normally solitary, pairing off only during the breeding season. However, when they mate, it is typically for life. Some owls share a nesting box through the winter, while others leave the nest and return during the next mating season to raise their brood.

Screech-owl courtship begins as early as the end of January and continues through early spring, when the female returns to the nest cavity and prepares to lay her eggs. She does not build a nest but instead uses her body to create a depression in debris in the cavity.

> ▶ **keeping a lookout**
>
> Most birds have eyes that are set toward the sides of the head. But an owl's large eyes, which help it look for prey at night, are fixed on the front of the bird's broad, flat face. To compensate for their forward-facing eyes, owls are capable of turning their heads almost completely upside down and nearly completely around so that they can always keep an eye on their surroundings.

The Family

After spending six to seven days inside the cavity, the female lays the first three or four eggs, one per day. She may lay two to three additional eggs, with shorter intervals in between. During this time the male brings food to his mate, although the female may leave the nest briefly at dusk or dawn.

It takes three to four weeks for the eggs to hatch, but the young chicks actually start peeping a day or two before hatching. Once they hatch, the female owl will either eat the eggshells or carry them away from the nest, to protect against parasites and predators.

During the first few weeks of life, the male owl provides food for the young family. He brings prey to the nest, where the female breaks the food into smaller pieces and feeds them to the babies. After four weeks, the young owls are encouraged to leave the nest. They remain dependent on their parents for another eight to ten weeks.

The Western Screech Owl

Western screech owls *(Otus kennicoottii)* share many traits with their eastern relatives, including ear tufts, large yellow eyes, and black bars and stripes (the latter are somewhat more pronounced in the western owls). These birds are usually gray, although brown ones are common in coastal areas of the Northwest. Slightly noisier than eastern screech owls, the western variety adds shrill whistles and trills to its vocabulary.

Western screech owls are found in lower elevations west of the Rocky Mountains, extending south into the western two-thirds of Mexico and north along the western coast of Canada.

▶ **the bird statistics**

The Western Screech Owl at a Glance

Length: 8½ inches

Diet: Insects, small birds, rodents, frogs, toads, and small reptiles

Habitat: Open areas, preferably near water. Also nests in wooded areas, mixed or deciduous forests, and orchards. Prefers oak trees

Region: Western third of the United States and Mexico

Breeding season: Begins in January

Nesting Requirements

Entrance hole: 2¾ inches in diameter, about 10 inches above the floor

Mounting height of box: 5 to 30 feet above ground.

Location of box: Near an open area, preferably in a deciduous tree near water. Owls should have a clear flight path to and from nest. Install predator guards.

Western Screech Owl Diet

Western screech owls are nocturnal predators, preying on everything from insects and small rodents to the occasional duck or pheasant. There have been accounts of western screech owls raiding pheasant farms, although these incidents are extremely rare.

Mating and Nesting Habits

The nesting habits of the western screech owl are similar to those of the eastern screech owl (see page 52). The western species is most commonly found nesting in oak or streamside woodlands.

The Family

Western screech owls generally have a brood (one per year) of two to five. Eggs may be laid as early as January and take three to four weeks to incubate. Young birds eat the insects and small rodents that their parents bring back to the nesting cavity. The chicks are dependent on their parents for several weeks.

> ▶ **clean with caution**
>
> The owl box will require less cleaning than other nesting boxes because the owls don't build a tight nest. Naturally, you don't want to open the box while owls are inside! If you have reason to believe that the box has been raided or abandoned during the breeding season, use extreme caution before opening it. Observe the box for several evenings, using binoculars if necessary, to be sure there is no activity. You should open the box only if you are completely certain that it is empty.

building a thatched screech owl house

Difficulty: Easy

Because of the size of the house and the birds that will occupy it, I use 2-inch galvanized deck screws to hold this project together.

Since owls do not build nests, I supply the box with leaves, wood chips, and bark before hanging it. Rather than assemble nesting materials, owls simply find a pile of leaves or rubble in a tree cavity and lay eggs on top of the pile. So create a small pile and see if they come to it.

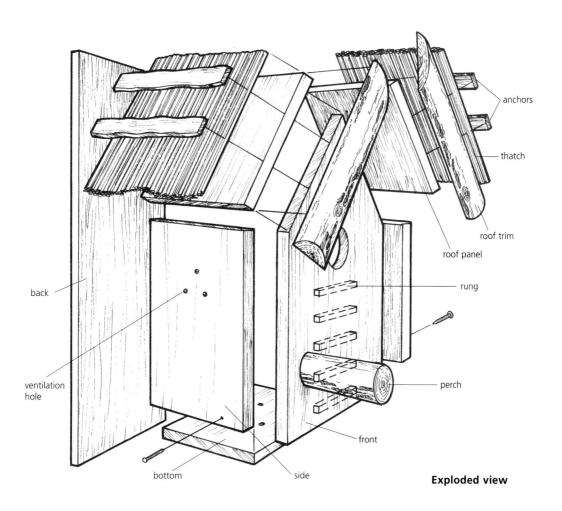

Exploded view

▶ what you'll need

▶ Parts
- Back: 1 x 12* of cedar or pine, 31" long
- Two sides: 1 x 10† of cedar or pine, 13½" long
- Front: 1 x 12* of cedar or pine, 19" long
- Roof panel: ¾" x 10¾" x 8" piece of plywood
- Roof panel: ¾" x 10¾" x 8½" piece of plywood
- Bottom: 1 x 10† of cedar or pine, 8¾" long
- Perch: 2½"-diameter branch, 5" long
- 6 rungs: ¼" x ¼" x 3"
- Door stop: ½" x 1" x 8"
- Roof trim: 3"-diameter branch, 15" long
- 4 roof anchors: ½" x 1½" x 10⅝" long
- 4 large handfuls of thatch, 15" long

*The true size of a 1 x 12 is ¾" x 11¼".
†The true size of a 1 x 10 is ¾" x 9¼".

▶ Materials
- 2" galvanized deck screws
- Brads
- 2" roofing nails
- Sandpaper (optional)
- Exterior wood glue
- Clear enamel spray gloss or waterproof lacquer

▶ Tools
- Circular saw
- Drill with screwdriver bit and ¼" and ½" drill bits
- Jigsaw
- Hammer
- Firewood wedge
- 3-lb. sledgehammer
- Pruning saw or handsaw
- Multipurpose scissors

Assembling the Owl House

1. Drill three ¼-inch-diameter mounting holes near the top and bottom of the back. Drill a ½-inch-diameter hole 10 inches from the top. ▼

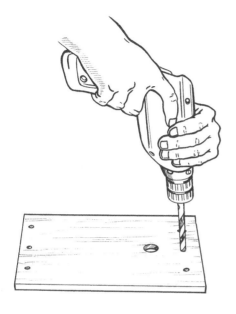

2. On the front, lay out a 3-inch-diameter entrance hole, centered 11 inches from the bottom. Lay out the angle of roof, following the dimensions in the drawing. Cut along the roof line with a jigsaw. To cut out the entrance hole, drill a ½-inch-diameter hole just inside the layout line. Slip the blade of the jigsaw through the hole, turn on the saw, and cut along the layout line. ▼

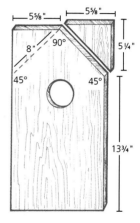

3. Place the perch off center, 4 to 6 inches below the front entrance. Attach it by driving two 2-inch galvanized deck screws from inside the house. ▼

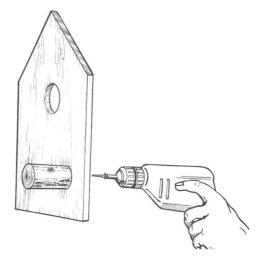

4. Young owls need an interior "ladder" to help them reach the entrance hole of the house. Attach the rungs to the inside front surface, spacing them roughly 1 inch apart. ▼

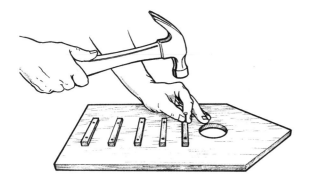

5. Drill three ¼-inch-diameter ventilation holes in each side piece. The floor is a trap door that pivots on two nails going through the sides and into the edge of the floor. Drill a ⅛-inch-diameter hole in each side for the nails, positioning the holes 2 inches in from the front and ½ inch up from the bottom. ▼

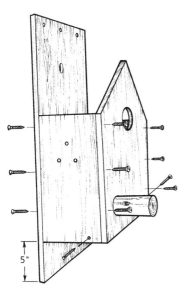

Screw the back to the sides, positioning the pieces as shown at left, below. Use three 2-inch galvanized deck screws on each side.

Screw the front to the sides.

Building the Trap Door

1. The bottom swings open like a trap door, and needs to be about ½ inch narrower and ½ inch shorter than the opening.

Measure the opening, and cut the bottom to size. Before you test-fit the bottom, temporarily drive a nail into it to act as a handle. Cut it to size; test-fit; and, if necessary, trim it with a jigsaw or sand the edges. When the fit is right, drive the nails. Position the doorstop on the back just below the door, and screw in place with two galvanized deck screws. (See photo on page 48 for placement.) Remove the nail that you installed as a handle, and drill six to eight ¼-inch-diameter drainage holes in the bottom. ▼

Attaching the Roof, Roof Trim, and Thatch

1. Glue and nail the roof to the box, overlapping the pieces as shown. Drive the nails through the roof and into the box, and drive a few nails through the back and into the roof. ▼

2. For the roof trim, split the 15-inch long, 3-inch-diameter branch in half, using a firewood wedge and a 3-pound sledgehammer. Cut one end of each at a 45-degree angle, using a handsaw or pruning saw. Nail both pieces to the house, with the angled edges meeting at the peak, as shown. ▼

3. Prepare the roof anchors by drilling three screw holes in each strip: one hole at each end (1½ inches in from the end) and one hole in the middle. Insert galvanized deck screws in the holes.

4. Place small handfuls of thatch on one face of the roof, patting it into place as you work. When the thatch is about an inch thick, place one of the wooden anchors on top of the thatch and screw it lightly in place, but not so tightly that you can't move the thatch back and forth.

Thatch the second side the same way you thatched the first.

5. Align the bottom edges of the thatch by tapping them with a wooden block, and then tap along the top of the roof to create an even peak. Tighten the screws holding the anchors, and firmly attach the remaining anchors. Trim the thatch with multipurpose scissors. (For complete directions, see Thatching a Roof on page 23.)

6. Spray the roof with a clear enamel spray gloss or waterproof lacquer. This gives visual appeal and will also help prevent the thatch from shifting over time.

bat box

Suspicions amongst thoughts are like bats amongst birds, they ever fly by twilight.
Francis Bacon

Bats are not birds. They're small mammals, and the only ones that can truly fly. In addition to eating a whopping number of mosquitoes, bats also eat other insect pests, such as gypsy moths, spruce budworm moths, tussock moths, and pine bark moths. In fact, all but three species of bats in the United States are insectivores and eat nearly their full body weight in insects each night.

Bats hunt for their prey not with their eyes, but by emitting a steady clicking sound that functions as sonar. The sound is generally too high for humans to hear, but the bat can hear it, and he listens as the sound bounces off nearby objects, telling him where they are. Although it sounds simple, the bat's ability to detect prey in flight is incredibly sophisticated.

A forked branch used for one side of the roof trim adds a personalized rustic touch to this box.

The Truth about Bats

The idea of providing lodging for bats may seem a bit odd, if not downright spooky. Bats are cloaked in unfortunate supernatural legends that raise fears about rabies and odd nesting habits. Before we go any further, let's take a look at the legends.

▶ **"Bats Get Caught in People's Hair."** Bats have no more interest in landing in your hair than you have in letting it happen. Given the bat's remarkable built-in sonar system, no self-respecting bat is going to mistake your wig for home — or anything else. If you have ever had an opportunity to watch these creatures hunt for insects, you know firsthand how accurate their sonar is, and how graceful and acrobatic they are in flight.

▶ **"Bats Have Rabies."** The fact is, you have a better chance of winning the state lottery or getting struck by lightning than of dying of a bat bite. While a small percentage (about one half of one percent) of bats get rabies, so do most types of mammals. But when a bat is infected with rabies, it becomes inactive, so the chance of a rabid bat flying into you, biting you, and giving you rabies is slim.

▶ **"Bats Suck Blood."** We can thank Dracula for this myth. While three species of vampire bat in Central America do suck the blood from sleeping cattle, that's about the extent of the bats' Dracula-esque behavior. The North American bats don't bother cattle. Their diet consists largely of mosquitoes. In fact, one bat eats about 600 to 1,000 mosquitoes and other insects per hour.

▶ **bats and temperature**
Ideally, a bat house should be fairly warm during the day, with temperatures ranging from 80 to 90°F. One way to help create the right temperature is to paint the outside of the bat house, using white to reflect light and heat and black or another dark color to absorb light and heat. In cooler areas, paint the house black and place it where it will receive early-morning sunlight. In warmer areas, paint the house white.

Types of Bats

There are two main species of bats that you will probably come across, although there are many more species in the United States. Check with a local wildlife or conservation organization to find out more about the types of bats that live in your area.

The Big Brown Bat

Despite its name, the big brown bat *(Eptesicus fuscus)* is only 3 to 4 inches long; its name is a result of a comparison with another common nighttime flyer, the little brown bat. The big brown bat is one of the most common bats in North America, ranging from Canada to parts of South America and the Caribbean.

Frequently spotted near wooded areas and water, this type of bat has shiny brown fur, although the bat is usually moving too quickly for you to see it. Big brown bats usually roost in trees and attics and are frequent bat-house residents.

The Little Brown Bat

Although its body measures only 3½" long (only slightly smaller than its "big" cousin), the little brown bat *(Lasiurus cinereus)* has a 10-inch wingspan. The little brown bat also occupies a more northern territory, ranging from Alaska, Canada, and the northern United States only as far south as Georgia. In the summer, little brown bats can be heard moving around and bickering during the day, perhaps in response to the heat. Like all bats, they emerge in the evening and head out to feed.

Hanging Your Bat House

Hang the bat house on a wall or in a tree, at least 15 feet above the ground. It is generally preferable to hang the box in a more secluded spot, so that the bats will not be disturbed by human activity during the day. Choose a slightly open location as well, especially one that receives morning sunlight and is located near water, gardens, or other insect-attracting areas. The path to the entrance should be free of any obstructions, such as tree branches, and should be sheltered from the wind.

Bats usually choose a roost in early spring, so hang the box in the fall or winter. Once they choose a roost, bats will return to it every year.

An important thing to remember in attracting bats to your home is the insect population. If the ground around the bat house is treated with pesticides, it's a bad place for a bat. Always keep in mind that you are bringing a wild animal into your community, so try to keep the environment as welcoming as possible.

There are almost 1,000 species of bats worldwide. Bats are the only mammals that truly fly.

building a thatched bat box

Difficulty: Moderate

The bat house is unlike other houses in this book. For starters, there's no entrance hole. Instead, the house is floorless, and the bats fly in from the bottom. Twigs, glued and nailed to the inside surfaces, give the bats places to hang from. The inside of the house has a divider, which is covered with more twigs, giving the bats even more places to hang.

A box this size holds up to 100 bats, most likely males or nonreproductive females. It often takes a year or more for bats to establish residence in a newly located bat house, so be patient.

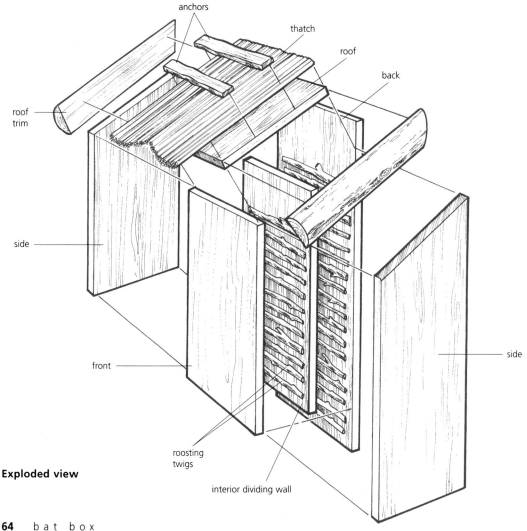

Exploded view

▶ What you'll need

▶ Parts
- Back: 1 x 12* of poplar or cedar, 26" long
- 2 sides: 1 x 6† of poplar or cedar, 20" long
- Front: 1 x 12* of poplar or cedar, 14¼" long
- Interior dividing wall: 1 x 12* of poplar or cedar, 12" long
- Roof: ⅝" x 11¼" x 8" piece of plywood
- About 20 roosting twigs: ½" diameter, 8–9" long
- Roof trim: 1 straight branch, 2–3" diameter, 12" long
- 2 roof anchors: ½" x 1½" x 9½" long

*The true size of a 1 x 12 is ¾" x 11¼".
†The true size of a 1 x 6 is ¾" x 5½".

▶ Materials
- Exterior wood glue
- 3d (3-penny) nails or 1¼" brads
- Black paint pen (optional)
- Black enamel spray paint (optional)
- About 2 to 3 handfuls of thatch, 12–13 inches long
- Clear enamel spray gloss or lacquer
- Screws for roof anchor

▶ Tools
- Circular saw
- Speed square
- Hammer
- Firewood wedge
- 3-lb. sledgehammer
- Pruning saw or handsaw
- Multipurpose scissors

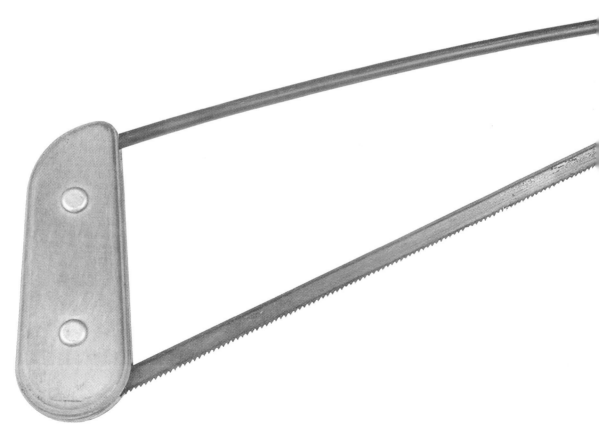

Angling the Sides

1. Cut a 45-degree angle along the top of each side piece, using a circular saw guided by a speed square. Sand off any rough edges that result. ▼

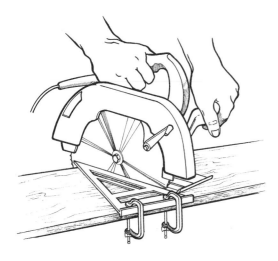

Assembling the Bat House

1. Using exterior wood glue, glue the sides to the back, aligning the bottom edges as you work. Drive a 3d nail through the top of each side into the back. Flip the bat house over, and drive a nail roughly every 3 inches through the back and into the sides. ▼

2. Give the bats a place to hang with several rows of twigs. Glue some twigs to the inside face of the front and back, spacing them about 1 inch apart. Then drive a nail into each end. Place similar rows of twigs on both sides of the interior dividing wall, for additional roosting space. ▼

3. Put the interior wall in the bat house, centering it from top to bottom. Nail the wall in place, driving a 3d nail through the sides every 2 to 3 inches. Glue and nail the front in place, driving 3d nails into the sides every 2 to 3 inches. ▸

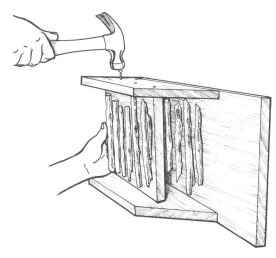

This bottom-up view reveals that two interior dividers have been added to this house to provide extra hanging rungs.

4. Glue and nail the roof to the bat house with 3d nails. ▼

Split a 2- to 3-inch-diameter, 12-inch-long branch in half with a firewood wedge and a 3-pound sledgehammer. Cut a 45-degree angle on one end of each with a pruning saw. Nail them in place at the edges of the roof, with about an inch extending above the roof, as shown in the exploded view on page 64.

Thatching the Roof and Hanging the Bat House

1. Prepare the roof anchors by drilling two screw holes in each strip, one hole at each end (1½" in from the end). Insert galvanized deck screws in the holes.

2. Apply thatch until the roof is approximately 1 inch thick. Screw the wooden strips loosely in place, then align the bottom edges of the thatch, and tighten the screws. Trim as desired. (For complete directions, see Thatching a Roof on page 23.)

3. Spray the front of the bat box, including the thatch, with a light coat of clear enamel gloss.

4. Drill two ⅜-inch-diameter holes near the top of the back for hanging the bat house. Drive screws through the holes to hang the house.

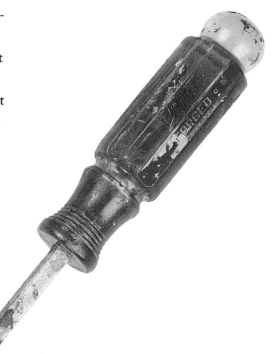

decorating the bat house

Full moons and flying bats add an air of mystery to this box. To add images to your house, see below.

Spray paints and paint pens are available in most craft and hobby stores.

Bat houses invite decoration. If you'd like to decorate yours, here are a couple of suggestions. Before you start, lay the bat house flat on your work space.

▶ Using a black paint pen, draw and color a bat on the box. Place the bat near one of the upper corners, leaving room (if you like) for a moon.

▶ To make a full moon, place the lid from a can of spray paint in one of the upper corners of the box, and spray a light mist of paint over that area. Draw two tiny M-shaped bats in the blank circle that results.

bat box **69**

bluebird box

The poetry of earth is never dead:
When all the birds are faint with the hot sun,
And hide in cooling trees, a voice will run
From hedge to hedge about the new-mown mead.
John Keats

Bluebirds, long considered symbols of happiness and good fortune, occupy a special place in our collective consciousness. Their colorful arrival announces springtime and sets the season in motion. Depending on the particular birds and your locale, a pair of bluebirds may continue to use the same nesting box for a few years.

There has been a resurgence in the bluebird population throughout North America since the mid-20th century, thanks in large part to people erecting bluebird boxes. Around the turn of the century, the bluebird population declined dramatically as a result of the loss of suitable nesting sites. Bluebirds are cavity nesters, but unlike woodpeckers, they can't drill their own holes. So when forest land was developed, bluebirds lost their natural nesting spots in tree cavities. The North American Bluebird Society, formed in 1978, started the effort to support bluebirds with boxes. You can help by putting up boxes in your neighborhood.

This rustic bluebird box adds charm and beauty to a weathered fencepost.

The Eastern Bluebird

Three types of bluebirds are found in the United States. Because of their widespread population, eastern bluebirds *(Sialia sialis)* are the most commonly recognized species. These birds occupy a large portion of North America, almost anywhere east of the Rocky Mountains. In areas where the climate is predominately warm, eastern bluebirds are year-round residents; otherwise they migrate to warmer climates during the winter.

You can recognize male eastern bluebirds immediately by the bright blue on their heads, tails, backs, and wings. They also have white bellies and chestnut-red patches on their sides. Females, in contrast, are fairly drab in appearance: They have light grayish-blue heads, blue tails, and dull brown backs. Female eastern bluebirds also have white rings around their eyes.

▸ **the bird statistics**

The Eastern Bluebird at a Glance

Length: 7 inches

Diet: Insects and spiders, fruit and berries (when other food is scarce)

Habitat: Open meadows, areas with a few trees for perching and hunting food in the grass below. Eastern bluebirds tend to avoid areas crowded with trees or people.

Range: Eastern and central United States, southeastern parts of Canada, and most of Mexico. Bluebirds in northern areas migrate south during the winter.

Breeding season: Begins with the first brood in mid- to late March to early May, depending on climate. One or two more broods follow.

Nesting Requirements

Entrance hole: About 1½ inches in diameter (anything larger invites European starlings and predators). Hole should be at least 6 inches above the floor.

Height of box above ground: 4½ to 6 feet. Make sure the box remains accessible for cleaning.

Location of box: Along the edge of an open area. Position the box so it faces south, to protect the bluebirds from wind and rain. If you live in an area where rain frequently comes up from the south, you may want to angle the box toward the southeast.

The Western Bluebird

The remaining western portion of the United States and Mexico is home to the western bluebird *(Sialia mexicana)*. Like the eastern bluebird, the western bluebird will remain in warmer climates year-round or migrate to and from northern areas as the seasons change.

The male western bluebird is the more colorful of the sexes. It is almost entirely blue, including its chin, throat, and tail. Like the eastern bluebird, it also has reddish patches on the breast and shoulders. The belly, however, is grayish blue, not white. Females are grayish brown on the head and back, with light blue on the wings and tail. Female western bluebirds also have a slight rust color on the breast and white eye rings.

▶ the bird statistics

The Western Bluebird at a Glance

Length: Maximum of 7 inches

Diet: Insects and spiders, fruit and berries (when other food is scarce)

Habitat: Open meadows, areas with a few trees to perch on while they hunt for food. Not as shy of crowded areas as eastern bluebirds

Range: Western United States and Mexico; northernmost tip of the range enters Canada

Breeding season: Begins in early April or May

Nesting Requirements

Entrance hole: About 1½ inches in diameter, although western bluebirds are not as threatened by European starlings as eastern bluebirds are. Hole should be at least 6 inches above the floor.

Height of box above ground: 5 feet or higher, but low enough to remain accessible for cleaning.

Location of box: Along the edge of, and facing, an open area. The box should be mounted so that it faces south, to protect bluebirds from wind and rain. If you live in an area where rain frequently comes up from the south, you may want to angle the box toward the southeast.

The Mountain Bluebird

The mountain bluebird *(Sialia currucoides)* occupies higher elevations in the central and western portions of the United States, as well as in western regions of Canada. This bluebird is similar to its eastern and western cousins, but its feeding and nesting habits are different to accommodate life at elevations above 7,000 feet. Among other things, the mountain bluebird tends to hover over meadows when foraging. This species also differs in that it rarely includes fruit or berries in its diet.

Male mountain bluebirds are bright blue on the head, back, wings, and tail, with a lighter blue tint along the throat and a grayish-white belly. Mountain bluebirds have no chestnut patches. The females are mostly grayish-brown, with pale blue wings and tail.

▶ the bird statistics

The Mountain Bluebird at a Glance

Length: 7 inches

Diet: Insects and spiders, very rarely fruit and berries

Habitat: More tolerant of trees and shrubs than eastern bluebirds but still require open land and meadows to hunt for and catch prey

Range: Central and western United States, western Canada (as far north as Alaska in the summer), and inland Mexico

Breeding season: Begins with the first brood in early April to late May, depending on climate and altitude

Nesting Requirements

Entrance hole: About 1½ inches in diameter (anything larger invites European starlings and predators). The hole should be at least 6 inches above the floor.

Height of box above ground: 4½ to 6 feet. Make sure the box remains accessible for cleaning.

Location of box: Along the edge of, and facing, an open area. The box should be mounted so that it faces south, to protect the bluebirds from wind and rain. If you live in an area where rain frequently comes up from the south, you may want to angle the box toward the southeast.

Bluebird Diet

The bluebird diet consists mostly of insects, usually the grasshoppers, flies, beetles, butterflies, and katydids that live in the meadows or open spaces that bluebirds inhabit. The diet offers a good explanation as to why bluebirds prefer sunny, open locations with lots of grass and a few trees.

Mating and Nesting Habits

Individual breeding habits depend on location and latitude. Eastern bluebirds generally begin the breeding season between mid-March and early April. Western and mountain bluebirds may wait until early April or May. To stake a nesting spot, the pair must compete with European starlings and house sparrows. Unfortunately, the bluebird usually loses a possible nesting place to one or the other of these birds. Although the fight for nesting sites was a problem for bluebird populations in the recent past, increasing use of birdhouses has brought the numbers back up.

Despite the differences in territory, all three bluebird species have similar nesting preferences. Bluebirds nest in and around open areas, preferring farmland, meadows, orchards, and other areas with scattered trees. The trees and shrubs provide perches for hunting for prey. They also give the male bluebird a place to sit while singing to claim his territory. Although western and mountain bluebirds will nest along forest edges, eastern bluebirds tend to avoid areas that are crowded with trees or people.

Once the birds select a suitable cavity, rotting tree, or birdhouse, the female begins building the nest. Males sometimes bring material to the nest, but their main responsibility is to protect their mate and the site until the nest is complete. Typically, the female completes the nest in four to six days. Once the nest is finished, she usually takes a few days off to rest before the real work of raising the family begins.

▶ keeping the competition away

Because competing house sparrows and European starlings nest before bluebirds in the spring, one way to prevent the "wrong" bird from using a nesting box is to check for and clean out a competitor's nest early in the season. Check for house sparrows frequently.

Another, less invasive method is to simply provide several nesting sites. Many birds are territorial within their species. A starling, for example, will take one of the birdhouses and defend the area against other starlings. The remaining houses are free for the bluebirds. For this purpose, space birdhouses 15 to 25 feet apart.

To keep house wrens from claiming a bluebird house, place it about 120 feet away from the cover of trees and shrubs.

The Family

Bluebirds lay four to six eggs, although there are sometimes as many as eight. The eggs are small, smooth, and shiny and range from solid white to pale blue. All of the eggs in one brood are the same color. An egg of a different color appearing in a nest indicates that another bird laid an egg for the mother bluebird to raise, a practice known as "egg dumping." Egg dumping is not uncommon, and the chick is usually raised as one of the brood, the parents being unaware of any differences.

Incubation generally lasts about two weeks. During that time and in the days immediately after the chicks hatch, males are responsible for providing food for the female. After the chicks "thermoregulate" (gain control of their own body temperatures), the mother will gradually leave the nest and begin to forage as well.

Sixteen to twenty days after hatching, the fledgling bluebirds will leave the nest for the first time. Although anxious to explore and learn more about the world around them, these young birds will remain dependent on their parents for a few more weeks. Halfway through the season, however, the female begins to work on a nest for her second brood. Although it is rare, the season's first brood will sometimes remain and help raise the second one.

Bluebird pairs usually return to the same location each year to nest. If last year's site is unavailable, they usually find a place nearby.

▶ share a house with a tufted titmouse

One bird you may want to encourage to nest in your bluebird house is the tufted titmouse. This small bird is about 6½ inches long and lives in the eastern half of the United States and along the eastern edge of Mexico. The titmouse is a gray bird with a white breast and belly, a tawny patch under the wings, and a gray crest. Its face is white around its bright, black eyes.

Titmice prefer nesting in natural cavities, but as cavities become scarce, the birds are slowly being forced to opt for birdhouses. Their greatest threat there is from house wrens, who destroy the eggs and nest. To give the tufted titmouse a leg up on the competition, place the birdhouse on a tree or fence post in partial shade, 5 to 15 feet above the ground. Place hair, short bits of thread, fur, and feathers near the nesting box to encourage titmice.

Titmice usually begin building their nests in late March to early April. The female lays five to six eggs within the following month. In warmer climates, titmice will raise two broods. In the north, where the warm season is shorter, they will only raise one.

building a thatched bluebird box

Difficulty: Difficult

When looking for places to hang your bluebird box, consider a "bluebird trail." An open area, such as a field or park where people frequently walk or jog, is perfect for one of these trails. After getting permission from the landowner, have a group of friends and neighbors each place a house along the path, every 10 yards or so. Hang the birdhouses in sunny, open spots, facing south, and make sure there are perches nearby. You can hang them on trees, fence posts, or poles that you supply. You may want to hang the boxes in pairs 15 to 25 feet apart to help accommodate tree swallows, which often compete with bluebirds but will not occupy two adjacent houses.

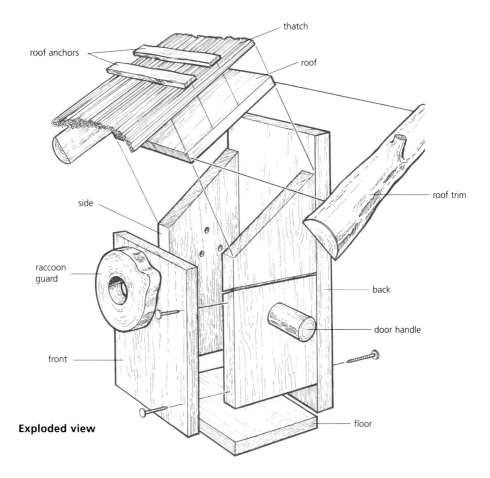

Exploded view

▶ what you'll need

▶ Parts
- Back: 1 x 8* of cedar or pine, 22" long
- 2 sides: 1 x 6† or cedar or pine, 14" long
- Front: 1 x 8* of cedar or pine, 10" long
- Roof: ¾" x 7¼" x 7¼" piece of plywood
- Floor: 1 x 8* of cedar or pine, cut to fit
- Raccoon guard: 4"-diameter log
- Door handle: 1"-diameter dowel or branch, 2" long
- Roof trim: 2"-diameter log, 12" long
- Roof anchors: make from 1½"-diameter log or dowel, 7" long

* The true size of a 1 x 8 is ¾" x 7¼".
† The true size of a 1 x 6 is ¾" x 5½".

▶ Materials
- Exterior wood glue
- 1½" aluminum or galvanized nails
- Three 1½" galvanized roofing nails to act as hinges for the door
- 1⅝" galvanized deck screws
- Sandpaper
- A bundle of straw or thatching material, about 4" inches in diameter and 20–40" long
- Clear enamel spray gloss or waterproof lacquer

▶ Tools
- Circular saw
- Speed square
- Hammer
- Drill with screwdriver bit and ⅛", ⅜", and 1½" drill bits
- Pruning saw or handsaw
- Firewood wedge
- 3-lb. sledgehammer
- Chisel
- Multipurpose scissors

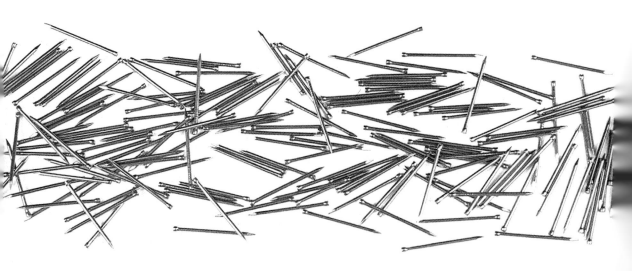

One of the best things about this house, besides the fact that it attracts bluebirds, is that it looks a lot harder to make than it really is. One tricky part of the assembly, however, is the door. Make sure it opens in the right direction when you fasten it to the sides.

In addition to giving you access to the inside of the house for cleaning, the door also lets you peek in on a nest. Give the house a little tap to make sure the mother bird isn't inside, then carefully open the door. Peeking in on the eggs, and later on the young chicks, is not only interesting but is also an opportunity to check on the nest and make sure the family is okay. If one of the chicks should die or an egg not hatch, remove it to prevent disease from affecting the other birds.

A ring of wood around the door helps keep raccoons from raiding the nest. Although a raccoon may be able to reach into the house, the limited size of the hole will keep it from taking anything out. The guard adds depth to the entrance hole, making it difficult for a raccoon to reach completely into the nest and grab an egg or a chick.

Ventilation and drainage holes allow air to circulate, prevent the bottom of the nest from becoming damp, and keep the birds from getting too hot in the summer.

The clean-out door opens out so you can monitor the nest without disturbing the birds or the eggs.

This rustic raccoon guard adds visual appeal as well as protection.

Cutting the Sides and Back

1. Lay out a 45-degree angle across the top of each side. Guide a circular saw along a speed square to cut out along the layout lines.

2. On one side piece, draw a line 4 inches up from the bottom edge of the peak. Set a circular saw to cut a 40-degree bevel, and cut the saw along the layout line with a speed square. The cutoff becomes the door that allows cleaning of the bird house. ▼

Assembling the Bluebird Box

1. Glue the 14-inch-long side to the back, aligning the top with the layout line on the back. Then attach the top piece of the side that you cut in two. Start with the shorter of the two pieces. Glue it in place with the peak on the layout line on the back. Look to see whether the bevel in the piece angles down toward the inside of the house. If not, swap sides, moving the left side to the right side, and vice versa. Nail the sides in place as shown. ▼

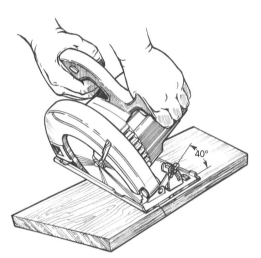

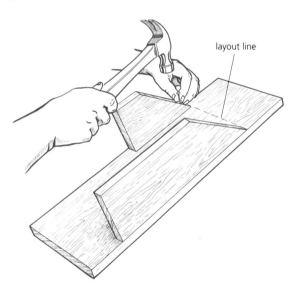

layout line

3. Draw a layout line across the back piece, 4 inches from the top. This will serve as a layout guide when you are placing the side pieces.

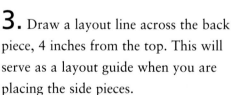

2. On the front, make a mark 4 inches from the top and centered from side to side. Drill a 1½-inch-diameter hole, centered on the mark. ▼

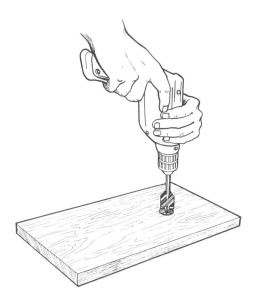

3. Glue the front in place on top of the sides, so that the upper edge is aligned with the place where the 45-degree slope begins on the sides. About 1 inch of the front will extend below the longer side piece. The entrance hole should be near the top. Nail the front in place. ▼

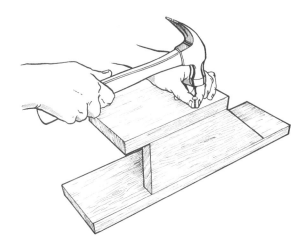

4. Glue and nail the plywood roof in place across the sloping angles of the side pieces. ▼

Installing the Floor, Clean-Out Door, and Raccoon Guard

1. Measure the opening between the back and front pieces, and cut the floor to fit in between the two pieces and against the bottom of the long side piece. Glue and nail it to the side, front, and back. ▼

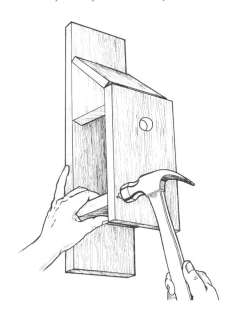

2. The door pivots on roofing nails driven through the front and back of the birdhouse. Drill ⅛-inch-diameter clearance holes, ½ inch from the edge and 2 inches from the bottom of each board. ▼

3. Drive a nail into the door to use as a temporary handle during installation. Put the door in its opening, and drive a 1½-inch roofing nail through each of the holes you drilled and into the door. ▼

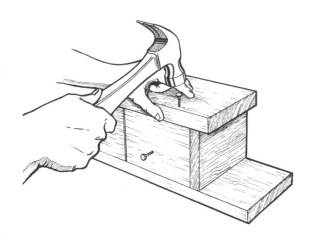

4. Replace the nail handle with the permanent handle made from a dowel or branch. Drill a ⅛-inch-diameter hole through the door and partway into the end of the dowel. Drive a deck screw through the back of the door and into the dowel.

5. For the raccoon guard, drill a 1½-inch-diameter hole, about 2 inches deep, in the end of a 4-inch-diameter log. With a handsaw or pruning saw, cut a 1-inch slice off the top of the log to get a donut-shaped disk. Glue and nail the disk over the entrance hole with 1½-inch brads. If you have trouble driving the brads, clip off the head of one brad, and use it as a drill bit to drill pilot holes through the log and into the house; then drive the brads through the pilot holes. ▼

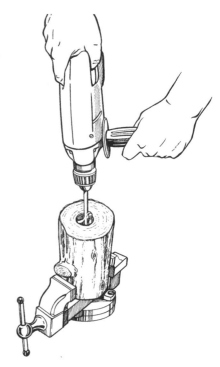

82   bluebird box

Adding the Roof Trim and Ventilation Holes

1. Split the 12-inch log in half with a firewood wedge and a 3-pound sledgehammer. Cut a 45-degree angle on one end of each half with a pruning saw or handsaw. Nail the pieces to the roof edges, leaving at least a 1½-inch lip to hold the thatch.

2. Drill three ⅜-inch-diameter ventilation holes in the solid side wall of the birdhouse. ▼

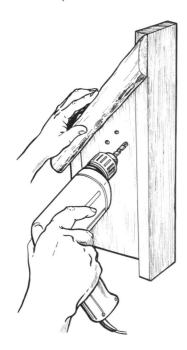

3. Drill a hole in the center of the floor for drainage, and sand the outside of the box.

Thatching the Roof

1. Split the 1½-inch-diameter dowel (or branch) in half lengthwise using a hammer and chisel. (Keep your hands well out of the path of the chisel.) Drill ⅛-inch-diameter holes about 1 inch from each end of the two pieces of dowel. Put a 1⅝-inch deck screw in each of the holes.

2. Place small handfuls of thatch on top of the roof, patting it into place as you work. When the thatch is about 1½ inches thick, place one of the wooden anchors on the top and screw it lightly in place, but not so tightly that you can't move the thatch back and forth.

3. Align the bottom edges of the thatch with the roof trim edges by tapping them with a wooden block. Tighten the screws holding the anchors, and firmly attach the remaining anchor. Trim the thatch, cutting a gentle curve in it where it overhangs the entrance. (For more complete directions, see Thatching a Roof on page 23.)

4. Spray the roof with a clear enamel spray gloss or waterproof lacquer. This is mostly for visual appeal but will also help prevent the thatch from shifting over time.

5. Drill a hole through the back of the box, about 8 inches down, and then hang the box on a nail.

nuthatch celtic cottage

I frequently tramped eight or ten miles through the deepest snow to keep an appointment with a beech-tree, or a yellow birch, or an old acquaintance among the pines.

Henry David Thoreau

The nuthatch is a familiar backyard feeder, most commonly recognized by its tendency to perch upside down on tree trunks, looking something like a small, wayward woodpecker. There are two common species of North American nuthatches, the white-breasted and red-breasted species.

The nuthatch feeds on many kinds of insects, including ants, beetles, flies, locusts, and spiders. It is a good bird to have around fruit trees since it can protect the trees from insect damage.

Nuthatches are often quite tame and are frequent vistors to feeders. With the nuthatch cottage in your yard, you're likely to see nuthatches often year-round.

Perched high among conifers near a clearing, meadow, or water, this charming cottage invites nuthatches to take up residence.

The White-Breasted Nuthatch

The white-breasted nuthatch *(Sitta carolinesis)* is the more common of the two species. These agreeable little birds are about 5 to 6 inches long and are usually seen in groups. They normally nest in tree cavities fairly high off the ground, which is important to remember when placing a nesting box for them. Try to find a location that is high, but not too inaccessible for the annual box cleaning.

Both the male and female white-breasted nuthatch have a dark blue-black back; a white chin, throat, breast, and belly; and reddish sides. The best way to distinguish between the two sexes is to note a subtle difference in the color of each bird's cap: black on males and dark gray on females. Young birds resemble their parents closely except for a greater amount of white on the breast. You can recognize a nuthatch by the slow "ank-ank-ank" of its call.

▶ the bird statistics

The White-Breasted Nuthatch at a Glance

Length: About 5¾ inches

Diet: Insects and invertebrates, seeds, nuts, and suet

Habitat: Prefers large, old, or dead trees and deciduous or mixed deciduous-coniferous forests. Usually nests along forest edges or in other open areas, near water, fields, parks, or orchards

Region: Most of the United States (especially the eastern half), small portions of southeastern and southwestern Canada, and Mexico

Breeding: Season starts in early April

Nesting Requirements

Entrance hole: 1¼ inches in diameter and about 5 inches above the floor

Height of box above ground: 12 to 20 feet (15 feet is a good average height)

Location of box: Nuthatches are more timid about birdhouses than other birds, so positioning is important. Try for a mature tree near a clearing, meadow, or water. To prevent house sparrows from taking over, place box away from buildings.

While these birds are at home in forested areas, they prefer to nest along forest edges and in more open areas. Unlike many other birds included in this book, nuthatches do not move with the seasons but instead remain in their territory throughout the year.

White-Breasted Nuthatch Diet

A large portion of the nuthatch diet comes from small invertebrates, such as ants, worms, larvae, and other treats that nuthatches find in tree bark. As the weather cools off and this food source begins to dwindle, nuthatches turn to other options, including acorns and seeds. Like chickadees, white-breasted nuthatches will also forage for animal fat, making frequent appearances at suet feeders during colder months.

Because nuthatches, like chickadees, will remain in an area year-round, they must take measures to save food for the leaner months. In the fall, nuthatches begin what's known as "scatterhoarding." They collect seeds and small nuts and carefully tuck them under the bark of trees, where they can be found and eaten later — if another bird hasn't found them in the meantime.

Mating and Nesting Habits

Pairs of white-breasted nuthatches work together to defend their breeding territory and will stay together for life. Even when the breeding season is over, they will stick it out through the winter and protect their territory.

The breeding season begins in the spring, when females start building nests in tree cavities, abandoned woodpecker holes, or birdhouses. The bottom of each nest has a base of earth and bark strips, which provides support for the egg cup. The cup is much softer and includes grass, hair, fur, and feathers. While the female builds, the male sometimes brings food. The two remain in contact during the process, often calling back and forth to each other.

At the beginning of the breeding season, nuthatches display a behavior known as "bill-sweeping," whereby they sweep their bills back and forth across the bark or wood near the entrance to their nest. It is not entirely clear why they do this, but because the birds are frequently seen bill-sweeping with insects crushed in their mouths, the theory is this: Insects often secrete toxic or foul-smelling substances, so it may be that nuthatches are trying to keep predators, such as squirrels and raccoons, from entering the nest.

The Family

After the nest is completed, usually in mid- to late April, the female nuthatch lays her eggs. The five to ten eggs in the clutch are glossy white, cream, or pinkish and have a covering of reddish to brownish spots, which may be more concentrated at the larger end. Once the clutch is complete, the female begins incubating the eggs, a process that lasts about 12 days.

During incubation and the first few days after the young hatch, the male flies back and forth to the nest to feed the female and their young. As the female gradually reduces the time spent in the nest, she begins feeding the young more frequently and begins helping the male clean waste from the nest. The young birds will stay in the nest for 14 to 17 days or longer. After fledging, they will still depend on their parents for several weeks.

Once the young birds leave the nest, the family shares the same territory until fall. In the fall, each of the offspring leaves the area to go in search of new territory. They stay in their new territories throughout the winter, then breed there the following spring. Depending on the availability of food, nuthatch parents may also relocate at the end of the breeding season. The white-breasted nuthatch has only one brood each year.

As you care for and clean your bird boxes over the years, you may find yourself collecting an array of finely built nests.

The Red-Breasted Nuthatch

One of the main distinctions between the red-breasted nuthatch *(Sitta canadensis)* and its white-breasted relative is territory. Red-breasted nuthatches prefer colder climates and higher elevations, and they have different diet and breeding habits as a result.

Of course, the other main difference between the two is coloring. Red-breasted nuthatches lack the white patches of their cousins. They have a slate-gray back and a rust-colored breast and belly. Both sexes have black caps and white eyebrows. Females have a slightly faded-looking appearance when compared to males, especially along the belly. Colors on juvenile birds are even more dulled and include speckled black markings on the eyebrow, chin, and sides of the head.

▶ the bird statistics

The Red-Breasted Nuthatch at a Glance

Length: Between 5 and 6 inches

Diet: Insects, invertebrates, seeds, nuts, and suet

Habitat: Prefers large, mature trees in coniferous or mixed coniferous-deciduous forests. Usually nests in aspen, spruce, or other cone-bearing trees. Also nests in birch, oak, cottonwood, and poplar

Region: Common in the western United States and mountain ranges along the eastern United States as well as in most of Canada. This bird prefers subalpine environments and is most common near mountains.

Breeding season: Starts late April to early May

Nesting Requirements

Entrance hole: 1 ¼ inches in diameter and about 5 inches above the floor

Height of box above ground: About 15 feet

Location of box: Similar to that for the white-breasted nuthatch. Birds will need encouragement, so choose a spot close to their natural habitat. Mixed coniferous-deciduous forests or an old hardy tree near a clearing is preferable.

The red-breasted nuthatch prefers cold climates, occupying areas in the northern and western United States and well into southern and middle Canada. Like white-breasted nuthatches, these birds occupy forests, although they prefer conifers to deciduous trees. They also seem to prefer slightly open forests and will nest in aspen, birch, cottonwood, poplar, and coniferous trees. Larger, older trees are also favored.

Red-Breasted Nuthatch Diet

Like its cousin, the red-breasted nuthatch forages up and down tree trunks in search of food within the bark. As invertebrates die off in colder weather, the red-breasted nuthatch looks to the seeds of cone-bearing trees, such as fir, pine, and spruce. They also frequent backyard sunflower and suet feeders in wooded areas.

Mating and Nesting Habits

Red-breasted nuthatches, like their white-breasted kin, are monogamous and stay together to defend a territory in the winter. In late April or early May, they excavate a workable cavity in a tree with their beaks, or choose a birdhouse. The soft nest inside the cavity is made of grass, roots, moss, hair, feathers, and fur.

To protect their nest from predators, a pair of red-breasted nuthatches will smear tree pitch around the entrance hole. The birds spend a great deal of time and energy on this process, smearing the pitch for several inches in every direction and continuing the process throughout the season. In order to keep this sticky mess outside the nest, red-breasted nuthatches frequently fly directly into the nest without alighting on the entrance hole first.

The Family

The female begins laying eggs once the nest is completed, and the finished clutch has four to seven eggs. The eggs are non-glossy white or pale pink with brown, reddish, or even lavender speckles. The female will incubate the eggs for approximately 12 days.

Both mother and father tend to the newly hatched brood for 18 to 21 days. Unfortunately, red-breasted nuthatches are elusive and little is known about how long parents spend raising the young birds — or even the number of broods they have.

Red-breasted nuthatches may move from place to place or remain in their territory, depending on the availability of fir, pine, and spruce seeds.

building a thatched nuthatch celtic cottage

Difficulty: Moderate

This little birdhouse is a favorite, with its white painted exterior and thatched roof capturing the look of a seaside cottage. In addition to providing a home for nuthatches, it will also attract wrens.

The interior of this cottage is divided in two by a wall. So this "duplex," with entrance holes on both ends, offers nesting spots for two pairs of nuthatches. The floor opens up to reveal both spaces for easy cleaning.

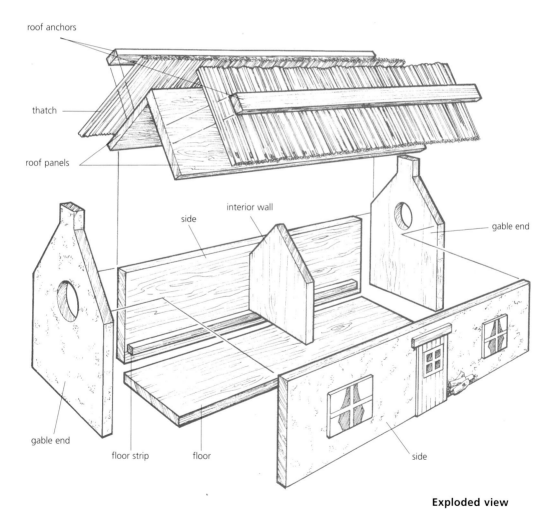

Exploded view

▶ what you'll need

▶ Parts
- 2 sides: ⅝" x 4½" x 12" piece of plywood
- Gable ends: ⅝" x 5" x 8" piece of plywood
- Roof panel: ⅝" x 3" x 12" piece of plywood
- Roof panel: ⅝" x 2½" x 12" piece of plywood
- Interior wall: ⅝" x 3¾" x 4¾" piece of plywood
- Floor: ⅝" x 3¾" x 12" piece of plywood
- 2 floor strips: ¼" x ¼" x 11" piece of plywood
- Roof anchors: 2"-diameter log, 12" long, split in two, or two ½" x 2" x 12" pieces of wood
- 2 windows: ¼" x 2" x 1½" piece of plywood
- Door: ¼" x 2" x 3" piece of plywood
- Lintel: ¼" x ¼" x 2½" piece of plywood
- Woodpile: ⅓"-diameter twigs, 1" to 1½" long
- 8 large handfuls of thatch, 8" long

▶ Materials
- ¼" brads
- ½" brads
- 4d (4-penny) nails
- 1" galvanized roofing nails
- 1⅝" galvanized deck screws
- Exterior wood glue
- 1 cup white latex primer
- ¼–½ cup sand
- Black latex spray paint
- Red or green latex spray paint
- Green, black, and white paint pens (available at hardware and craft stores)
- Screw for anchor

▶ Tools
- Circular saw
- Utility knife
- Jigsaw
- Hammer
- Drill with 1½" and ¼" bits
- Sandpaper
- ¾" piece of dowel
- Paintbrush
- Firewood wedge
- Multipurpose scissors

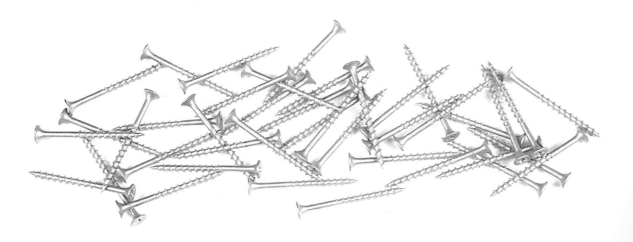

Assembling the Ends and Sides

1. To shape the gable end pieces, make a full-sized template using the drawing below as a guide, lay it on your piece of wood, and trace around it with a utility knife to lay out the chimney. Cut along the layout lines with a jigsaw to shape the roof sides and chimney. ▼

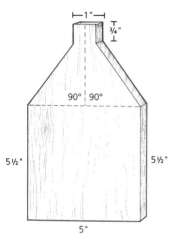

2. Glue and nail the gable ends to the sides. Starting with one corner of the house, apply glue to edge of side and gable. Support one of the pieces on a box or piece of scrap, then nail parts together with 4d nails. Repeat on the remaining corners. ▼

Attatching the Roof and Interior Wall

1. Make sure the roof pieces fit between the chimneys, and adjust the length if necessary. Nail the 2½-inch roof piece in place first. Nail the 3-inch piece in place so that it overlaps the smaller piece, as shown. ▼

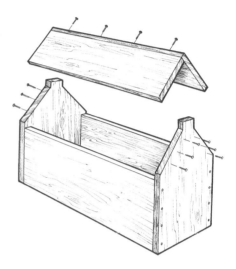

2. On the interior wall piece, cut 1½-inch pieces off the corners with a jigsaw so that the piece will nest underneath the roof. Slip the wall in place and measure to make sure there is 1 inch between the bottom of this wall and the bottoms of the adjoining sides. If not, trim the wall with a jigsaw until the spacing is correct. Glue and nail the interior wall in place. ▼

Finishing the Exterior

1. Drill the entrance holes. Drill a 1½-inch-diameter entrance hole in both ends of the birdhouse, 4½ inches from the bottom. Sand the edges of the entrance holes with a piece of sandpaper wrapped around a thick piece of dowel. Sand down all rough pieces and edges on the outside. ▼

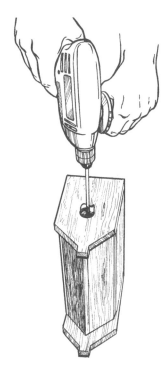

Painting the Birdhouse

1. The house is stuccoed with a mixture of white latex paint and sand. Mix the sand into the paint, a little at a time, to create a thick paste that you dab onto the house with a brush. Cover the birdhouse with this mixture, and allow it to dry thoroughly for at least 24 hours. ▶

painting doors & windows

▶ Paint the windows and all sides of the lintel with black latex spray paint. To ensure a deep enough color, apply two coats.

▶ Paint the door with your choice of green or red, and let door and windows dry. Apply additional coats as needed.

▶ To create the window details, use a white paint pen, and draw a cross to create the "panes" on the windows. Draw curtains inside the windows, using a green paint pen. Be careful not to color over the white lines, so the curtains appear to be inside the house.

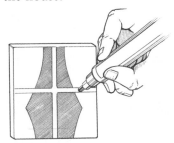

▶ Use the black and white paint pens to draw a window and doorknob.

Applying the Details

1. Attach the door, lintel (piece that extends across the top of the door), and windows to the front of the home with ¼-inch brads.

2. To create a mini-woodpile, position the twigs next to the door and nail them onto the house, using ½-inch brads. ▼

2. Drill five ⅛-inch-diameter mounting holes in the floor. Put the floor in the house, and drill a ⅛-inch-diameter hole through each side and into the side of the floor piece. Slip nails into the holes to keep the two pieces (house and floor) attached. ▼

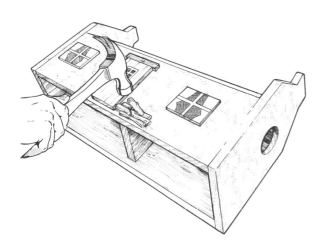

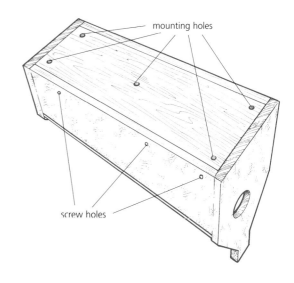

Installing the Floor

1. Nail two floor strips ¾ inch from the bottom of the sides, as shown in the exploded view. The floor will rest against these when assembly is complete.

3. When you're ready to install the birdhouse, remove the floor from the house and attach to the top of a tall post or mounting shelf first, then set the house on top of it. Finally, mount the post.

Thatching the Roof

1. Prepare the roof anchors by drilling three screw holes in each strip: one hole at each end (1½ inches in from the end) and one hole in the middle. Insert galvanized deck screws in the holes.

2. Place small handfuls of thatch on one face of the roof, patting it in place as you work. When the thatch is about 1 inch thick, place one of the wooden anchors on the top of the thatch and screw it lightly in place, but not so tightly that you can't move the thatch back and forth.

3. Using a hammer or piece of wood, lightly tap the top edges of thatch to line it up with the wooden roof.

4. Thatch the second side the way you thatched the first. (For complete directions, see Thatching a Roof on page 23.)

5. Align the bottom edges of the thatch by tapping them with a wooden block, allowing an overhang of 1 to 2 inches below the roof edge. Then tap along the top of the roof to create an even peak. Tighten the screws holding the anchors. Trim the bottom edge of thatch to desired shape and length with multi-purpose scissors.

6. Spray the roof with a clear enamel spray gloss or waterproof lacquer to help prevent the thatch from shifting over time.

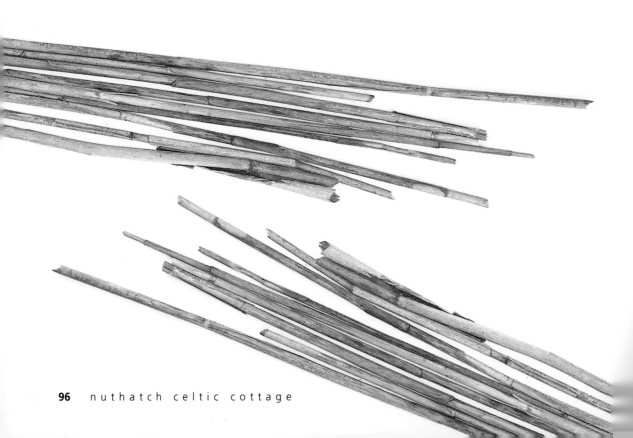

Have fun with the "home decorating" details on this cottage! You can even paint the doors and windows to match the colors of your own house.

tudor wren house

*I believe a leaf of grass is no less
than the journey-work of the stars,
And the pismire is equally perfect,
and a grain of sand, and the egg of the wren . . .*
Walt Whitman

The house wren is one of the most charming birds you can invite into your yard. These cheerful little brown birds are friendly, entertaining fellows who soon learn how to win bits of food from their human neighbors.

Unfortunately, they have also come face to face with some controversy. In recent years, nesting boxes may have contributed to an increase in wren populations, putting nesting sites at a premium. As a result, wrens fiercely protect any sites they can find, sabotaging the nests of other birds and piercing eggs when necessary. If wrens are frequent visitors to your yard, hanging boxes for them may help relieve the competition for nesting sites!

The body of this house is constructed of plywood and covered with a stucco-like finish.

The House Wren

the bird statistics

The House Wren at a Glance

Length: Between 4½ and 5¼ inches

Diet: Mostly invertebrates taken from trees and nearby foliage, including millipedes, spiders, snails, caterpillars, grasshoppers, ants, and beetles

Habitat: Areas with a variety of vegetation as well as more open areas. Suitable wren habitats range from swamps and forest edges to farmland and meadows. Wrens seem to appreciate a nest with a downhill view.

Region: Northern two-thirds of the United States and parts of southern Canada

Breeding season: Begins late April to early May, depending on climate

Nesting Requirements

Entrance hole: 1 inch in diameter. Hole should be 6 inches above floor.

Height of box above ground: 4½ to 10 feet. Install box so that it is accessible for cleaning.

The house wren *(Troglodytes aedon)* is a very common bird. In the winter, birds in the northern areas of the United States will migrate to southern areas of the country and into parts of Mexico. Their nesting habits are similar to those of the bluebird: they prefer fairly open areas in order to forage for the insects that make up most of their diet. These little birds are not quite as selective as bluebirds, however, and will nest in forest edges, parks, and farmland as well as swamps and shrub-covered areas.

Both male and female wrens have brownish gray upperparts, buff underparts, and brown "eyebrows." But perhaps their most distinguishing feature is a vertical tail, most often noticed on males as they whistle and twitter in their efforts to entice females to nest.

House Wren Diet

While wrens tend to find an ample food supply in and around trees, they do migrate south in the winter to continue foraging for insects.

Mating and Nesting Habits

Males arrive at the start of the breeding season and begin to stake out their territory. Once he is established, the male wren builds a series of "dummy nests" around his territory. He tries to build a nest in any and all available nesting spots, perhaps in an effort to secure as many broods as possible. The male's nests consist of a pile of small sticks, sometimes a couple of hundred, placed in a cavity, such as a tree, a birdhouse, or even an old hat.

Female wrens arrive later and, after selecting a mate, may choose one of the nests in the mate's territory or build one of her own. In building her nest, or egg cup, the female constructs a small, cup-shaped bed out of feathers, hair, wool, cocoons, bark, moss, and other soft materials, behind the pile of sticks.

A male may choose one mate for the first part of the breeding season and then move on to a different female for the remainder of the season, or he may mate with two females throughout the entire season. Wrens are very territorial, and the male may destroy the eggs of other birds that build nests within his territory.

Males return to the same territory each year, but females select different mates each year. Wrens usually raise two broods each season.

▶ housecleaning

Because wrens raise two broods per season, it helps to keep an eye on the box, cleaning it out when the family leaves. Sparrows and wrens will fight over an established nesting box. I've had to clean out at least one nest once the battle was over. Shortly thereafter, the wrens returned to the clean box to try again, this time without interference.

The Family

Like most birds, the female wren lays one egg per day until she completes the clutch, which usually consists of six to eight eggs. The eggs are glossy white, pinkish, or light buff and are covered with varying shades of brown specks. The female begins to incubate the eggs as the clutch reaches completion, with incubation lasting from 13 to 15 days.

The chicks are cared for by both parents and fledge after two weeks, although the parents continue to care for them for another two weeks. As the first half of the breeding season draws to a close, the female will begin to renest. If the young are still dependent on their parents as the female prepares for her second brood, the male will finish raising them.

building a thatched tudor wren house

Difficulty: Moderate

Perhaps as evidence of their willingness to nest almost anywhere, wrens are the only birds that will nest in a hanging box. The design of this house takes advantage of their willingness, leaving more stable sites available to other cavity nesters.

To give this cottage a more rustic appearance, I mixed sand with paint and painted on the stucco-like paste. I recommend two coats or more and depending on the weather, the paint may need a few hours or an entire day to dry thoroughly between coats. I'd suggest allowing a few days to build this house.

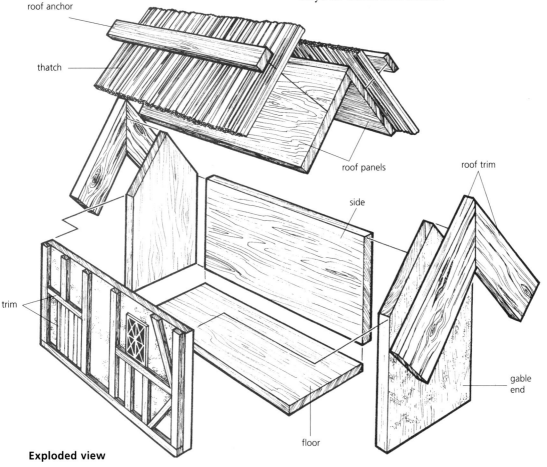

Exploded view

▶ What you'll need

▶ Parts
- 2 gable ends: ¾" x 5" x 8" piece of plywood
- 2 sides: ¾" x 4½" x 8" piece of plywood
- Roof panel: ¾" x 9½" x 3⅞" piece of plywood
- Roof panel: ¾" x 9½" x 4⅝" piece of plywood
- Floor: ¾" x 7⅞" x 3½" piece of plywood
- Door: ¼" x 2" x 3" piece of plywood
- 2 windows: ¼" x 2" x 1½" piece of plywood
- 2 shutters: ¼" x ½" x 3" piece of plywood
- 4 roof trim pieces: make from 1 x 3* of cedar or pine, about 30" long
- 4–5 trim pieces: ¼" x ¼" x 36" wood strips, available at hobby stores
- 2 roof anchors: 9½" x 1" x ½" thick
- Doorstop: scrap of plywood, at least 1" x 1"

*The true size of a 1 x 3 is ¾" x 2½".

▶ Materials
- Exterior wood glue
- 3d (3-penny) nails
- 4d (4-penny) nails
- 1" brads
- Galvanized roofing nails
- Sandpaper
- 1½ cups washed sand (although any sand works)
- 1 cup white exterior latex primer
- Green, red, or yellow enamel spray paint
- Black enamel spray paint
- Double-sided carpet tape
- White paint pen (available at hardware and craft stores)
- Green, red, or yellow paint pen
- 1" wire fence staple
- 1- to 2-foot length of chain for hanging the birdhouse
- Screw for anchor
- 2 big handfuls of thatch

▶ Tools
- Circular saw and speed square, or jigsaw
- Hammer
- Drill with 1½", ⅛", and ⅜" drill bits
- Utility knife
- 2" paintbrush
- Multipurpose scissors

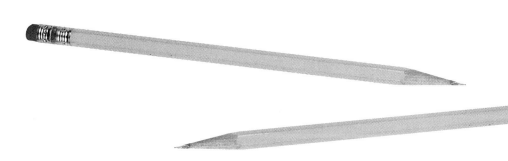

Preparing the Sides and Assembling the Wren House

1. Cut off two corners on each side, as shown, to create the gable end of the house. Make the cut with either a jigsaw or circular saw. If you make the cut with a circular saw, guide it against a speed square, and rest the saw on the gable, not the scrap. ▼

2. To attach the walls, start by placing one of the gable ends on edge on a flat surface. Apply glue to the exposed edge, and then place a side over it, supporting it with some scrap wood or a small box. Nail the two together with 4d nails.

Glue and nail the side to the other gable end, and then glue and nail the remaining side in place. ▼

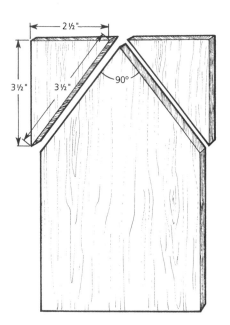

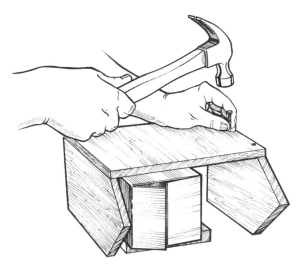

3. To assemble and attach the roof, put the 3⅞-inch piece on top of the house, and nail it in place with 4d nails. Put the 4⅝-inch piece on the roof, and nail it in place with 4d nails. There will be gaps between the roof and the front and back pieces, which will help provide ventilation. ▼

4. For the entrance hole, measure 4½ inches up from the bottom of the birdhouse. Make a mark where the center of the hole should be, then drill a 1½ inch-hole on the mark. Smooth out the hole with sandpaper. ▼

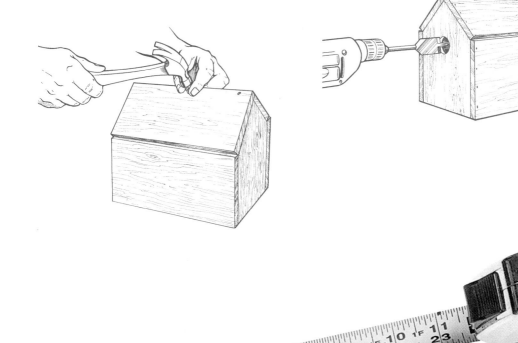

Painting the Box and Trim

1. Pour a cup of white latex primer into a small container, add a small amount of sand, and stir. Continue adding sand and stirring until you have a very thick paste.

Dab a fairly thick coating of the mixture on the outside walls of the house. Place the house outside or in a well-ventilated area, and let it dry for at least 24 hours. ▼

3. Sand smooth the door and window pieces. Carve two V-grooves into the door with a utility knife. Cut away from your body, and stick the piece to the table with double-sided carpet tape, so you won't cut your hand. When you're done with the doors, cut a single groove into each of the shutters, using the same technique. ▼

2. For the roof trim, cut the end of a 1 x 3 at a 45-degree angle. Then cut a length measuring 7 inches from the top of the cut. Cut three more trim pieces the same size and shape. Sand the wood, and then spray a thin coat of green, red, or yellow spray paint over both pieces, leaving the grain partially visible underneath.

Spray the door and shutters the same color as the roof trim, applying a coat thick enough to partially obscure the grain. Apply two coats of black spray paint to the windows and to all sides of the trim strips, obscuring the grain completely.

4. Draw diagonal lines ¼ inch apart on the window sections with a white paint pen, to make the windows look leaded.

Installing the Floor

1. The floor swings open, like a trap door, to allow you access to the house. Drill 1/8-inch-diameter holes in the front and back of the house for the nails that act as pivots for the floor. Position each hole 1/4 inch from the bottom and 2 inches from the end with the entrance hole.

Test-fit the plywood floor inside the bottom of the birdhouse, and trim as necessary, allowing about 1/8" all around. Drive a 3d nail into the floor to use as a temporary handle while working. The trap door should open and close easily. Put the floor in place, and hammer a galvanized roofing nail into each of the holes you drilled into the sides. Drive the nail tightly against the side, so that you can put trim over it later. Remove the 3d nail handle from the floor. ▼

Applying the Trim

1. When the paint is dry, put the roof trim on the gable end, so that about 1 inch of trim extends above the roof. Nail the trim in place with 4d nails. ▼

2. Trim the bottom, front, back, and sides. Apply trim around the bottom of the birdhouse, cutting it to fit. Color any exposed ends with a matching-color paint pen. Using 1-inch brads, nail vertical trim to the corners at the front, back, and sides. ▼

3. Attach the door to the front of the birdhouse, resting it on the trim below it. Apply a piece of trim above the door, and another piece to the left.

Place a window on the house next to the door or just below the roof. If desired, add pieces of trim over the top only to set it apart.

4. To imitate timber construction, cut some of the trim pieces at a 45-degree angle and run them diagonally between the horizontal and vertical "timbers." Trim the back, adding windows or doors as desired.

5. The shutters go on the side opposite the entrance hole. Before attaching them, hammer a 1-inch brad into the side where the two closed shutters will meet. Attach the shutters with a few brads or 3d nails, using the first nail you drove to keep the shutters from laying flat against the wood. The shutters will look as though they are slightly open and will add to the charm of the finished product. ▼

Completing the Bottom

1. To keep the bottom from closing too far, install a small scrap of plywood to act as a doorstop inside the birdhouse. Using 1-inch brads, nail the piece to the wall opposite the entrance, ¾ inch from bottom edge so the closed door will rest against it.

2. Drill a ⅛-inch-diameter hole in the trim on the front of the birdhouse, 2 inches in from the shuttered side. Insert a 4d nail into this hole to lock the bottom shut. The nail should remove easily for cleaning. ▼

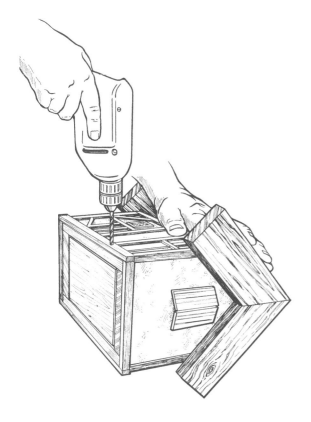

Attaching the Chain and Applying the Thatch

1. Nail one side of a 1" wire fence staple into the center of the rooftop. Attach the desired length of chain, and securely hammer both sides of the staple into the roof.

2. Prepare the roof anchors by drilling three screw holes in each strip: one hole at each end (1½ inches in from the end) and one hole in the middle. Insert galvanized deck screws in the holes.

3. Place small handfuls of thatch on one face of the roof, patting it in place as you work. When the thatch is about an inch thick, place one of the wooden anchors on the top of the thatch and screw it lightly in place, but not so tightly that you can't move the thatch back and forth. Tap lightly along the top of the roof to even out the thatch.

Thatch the second side the way you thatched the first, and attach anchor.

4. Align the bottom edges of the thatch on both sides by tapping them with a wooden block, and then tap along the

The details are what make this house so charming, and fun to make.

top of the roof to create an even peak. Tighten the screws holding the anchors, and firmly attach the remaining anchor. Trim the thatch with multipurpose scissors. (For complete directions, see Thatching a Roof on page 23.)

5. Spray the roof with a clear enamel spray gloss or waterproof lacquer. This is mostly for visual appeal but will also help prevent the thatch from shifting over time.

MASTER THATCHER tips

The Tudor look of this birdhouse comes from the pieces of brown trim that decorate it. Use a little (or a lot of) artistic license when you apply it. The door, windows, and other decorations on the outside of the house are only one approach, and the measurements given are suggestions. If you like, apply trim around the doors and windows. Diagonal pieces connecting horizontal and vertical trim add to the effect.

If you'd like to experiment with trim as you go, cut several extra trim pieces about 20 inches long, and then cut them to fit as you work.

If you're experimenting, don't stop with the trim. Try using twigs to make a small "woodpile" outside the house. Make a small bench that doubles as a perch. Paint flowerpots or window boxes on the house, or decorate with a particular theme in mind, such as Christmas.

butterfly house

I do not know whether I was then a man dreaming I was a butterfly, or whether I am now a butterfly dreaming I am a man.

Chuang Tse

The term "butterfly house" is a bit misleading. Although these decorative garden structures are gaining popularity in America, the fact is that butterflies do not nest in a house. So what exactly do these boxes do?

A butterfly box provides shelter for overwintering butterfly species — not a home during the sunny spring and summer months. While their popularity might suggest that the boxes attract many winged wonders, there are actually not a whole lot of over-wintering butterflies. You might decide that this box provides an attractive garden accent, even if it doesn't house any butterflies.

For best results, mount your house in a wooded area near logs or bark, where butterflies typically hibernate.

▶ the butterfly garden

The concept for this garden is grounded in the fact that butterflies are drawn to colorful flowers. The butterflies eat the nectar from some plants and lay their eggs on other plants, which become a food source for their caterpillars.

Plant the garden in a sunny, open place, providing rocks or other places for butterflies to shelter when the weather becomes cold or rainy. All of the flowers listed here attract butterflies. Check with a local nursery to find out which ones grow well in your area, and select varieties that will ensure that your garden will remain in bloom throughout the season.

- African lily (Agapanthus africanus)
- Astilbe (Astilbe spp.)
- Bee balm (Monarda didyma)
- Butterfly bush (Buddleia spp.)
- Candy tuft (Iberis sempervirens)
- Coneflower (preferably purple) (Echinacea spp.)
- Cupid's dart (Catanache caerula)
- French marigold (Tagetes patula)
- Hollyhock (pink or purple) (Acea rosea)
- Lantana (Lantana spp.)
- Leopard flower (Belamcanda chinenis)
- Various annual (flowering) salvia (Salvia splendens, S. "Victoria Blue," and S. farinacea)

Hibernating Butterflies

A butterfly begins its life cycle as an egg, after which it enters the larval stage as a caterpillar, then the pupa stage as a chrysalis, finally emerging as an adult. While most butterfly life cycles work so that the insect hibernates during one of the first three phases, a few species overwinter by hibernating as adults. These are the species you want to attract to a butterfly box, and they include the mourning cloak, Milbert's tortoiseshell, zephyr comma, question mark, and American snout.

When choosing a location for the butterfly house, keep in mind that this structure is for hibernating butterflies. Place it in a dark, sheltered place, where butterflies would typically go to hibernate (they like to be under logs or in tree bark, for example). Put the box in a woody section of your yard on trees or near vines that butterflies like. These include elm, willow, and poplar trees and hop vines. The mounting height of the box doesn't seem to matter, nor does the direction the box faces.

Remember to check the inside of the box from time to time for wasp nests or spider webs. One way to avoid such unwanted guests is to wait until later in the season to place your box outside. When you set out the box, remember to put a branch or a piece of bark inside for butterflies to "roost" on. A few thin branches from one of the suggested plants may provide more encouragement.

▶ life cycle of the mourning cloak

The mourning cloak is one of the North American hibernating butterflies that, if you're lucky, may inhabit your butterfly box. It begins life as an ebony caterpillar. The chrysalis that the caterpillar forms varies from whitish tan to bluish black with pink-tipped bumps. The adult butterfly that emerges is unmistakable in its coloring — brownish-maroon wings are framed with a cream border. Mud, fruit, and tree sap lure the butterfly out of hibernation in the spring.

caterpillar

chrysalis

adult

butterfly house **115**

building a butterfly house

Difficulty: Moderate

Many people choose to decorate the box on the outside, sometimes with painted flowers. While this probably doesn't have much of an effect in convincing butterflies to hibernate there, it doesn't hurt, either, and can add a nice touch to your yard.

Butterflies enter the box through one of a series of slits on the front, which serve as doors. These slits should be evenly dispersed in a pattern of your choosing, taking into consideration any decorations you've planned.

▶ what you'll need

▶ Parts

- Note: Use yellow pine (cedar repels insects) for entire project.
- Front: 1 x 8* of yellow pine, 20" long
- Back: 1 x 8* of yellow pine, 20" long
- 2 sides: ¾" x 4" x 16¼" piece of yellow pine
- Roof panel: ¾" x 5" x 5" piece of plywood
- Roof panel: ¾" x 5" x 5¾" piece of plywood
- Trap door: ¾" x 3½" x 5½" piece of yellow pine
- 4 roof trim pieces: make from 1 x 3† piece of yellow pine, about 30" long
- 4 roof anchors: made from two 1"-diameter logs, 10" long, to secure thatch

*The true size of a 1 x 8 is ¾" x 7¼".
†The true size of a 1 x 3 is ¾" x 2½".

▶ Materials

- Exterior wood glue
- 4d (4-penny) galvanized nails
- 1½" brads
- 2" galvanized screws
- 3 handfuls of thatch, 10" long
- Clear enamel spray gloss

▶ Tools

- Jigsaw
- Drill with ⅛" and ¼" drill bits
- Hammer
- Circular saw
- Speed square
- Multipurpose scissors
- Firewood wedge
- 3-lb. sledgehammer
- Pruning saw or handsaw

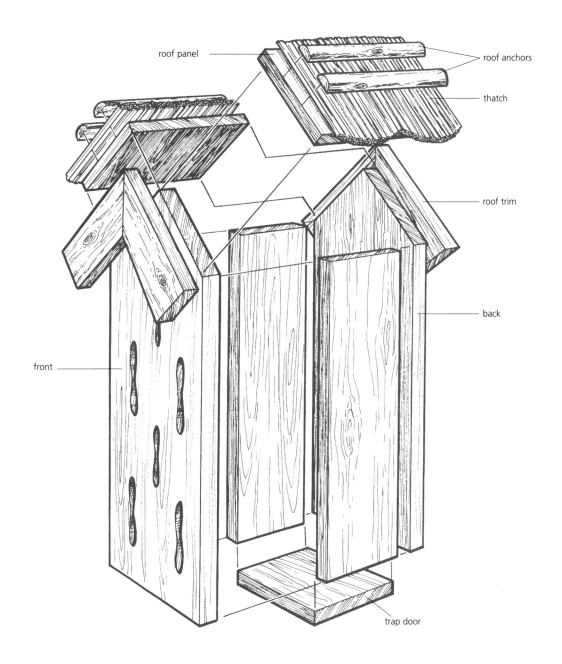

Exploded view

Assembling the Butterfly House

1. Lay out the peak of the roof on the front and back pieces, following the exploded view. Draw five or six lines on the front, each 4 inches long, marking the center of the entrance slits. ▼

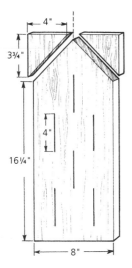

2. Cut along the layout lines for the roof with a jigsaw. Begin the cuts for the entrance slots by drilling a ¼-inch-diameter hole through one end of each layout line. Put a jigsaw blade through one of the holes, and cut out an irregularly shaped opening that is no more than ½ inch wide. Repeat until you've cut out all the openings. ▼

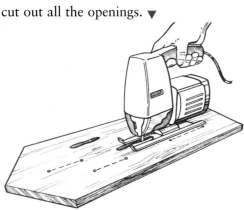

3. Apply glue to one edge of each side piece. Align the bottom edge of one side with the bottom edge of the front, and then nail the pieces together with 1½-inch brads. Attach the other side, and then attach the back, being careful to align the bottom edges. ▼

Attaching the Roof and Edging

1. Glue and nail the roof in place, nailing the larger piece on top of the smaller, as shown, to create a symmetrical roof. ▼

2. For the roof trim, cut the end of a 1 x 3 at a 45-degree angle from the top of the cut. Then cut a length measuring 7 inches. Cut three more trim pieces the same size and shape. Nail the trim pieces to the roof, so that each forms a 1-inch lip that helps hold the thatch in place. ▼

Preparing and Installing the Trap Door

1. This house has a trap door at the bottom to allow you access as needed. Instead of hinges, the door pivots on nails that run through the front and back and into the edge of the door. Drill a ⅛-inch hole for the pivot at the bottom of the front and back, positioning it 2 inches from the right side edge and ¼ inch from the bottom edge. ▼

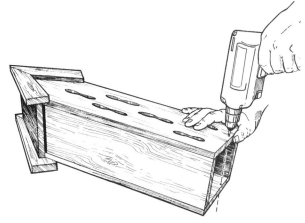

2. Drive a nail into the bottom of the trap door as a temporary handle, and then place the trap door into the bottom of the house. Trim the door as necessary to allow 1/8" clearance all around. Drive a 4d galvanized nail through the drilled holes and into the trap door. Make a latch for the door by hammering a nail partway into the bottom edge of the left side and then bending the nail over the door. ▼

Thatching the Roof and Mounting the House

1. Split the two 1-inch-diameter logs (or dowels) in half lengthwise using a hammer and chisel. (Keep your hands well out of the path of the chisel.) Drill 1/8-inch-diameter holes about 1 inch from each end of the four pieces. Put a 1 5/8-inch deck screw in each of the holes.

2. Place small handfuls of thatch, in layers, on one face of the roof, patting it in place as you work. When the thatch is about an inch thick, place one of the wooden anchors on the top of the thatch and screw it lightly in place, but not so tightly that you can't move the thatch back and forth. Tap the ends of the thatch with a piece of wood to align the ends with the peak of the roof. Repeat on the other side. With multipurpose scissors, trim the overhanging ends at the eaves. (For complete directions, see Thatching a Roof on page 23.)

3. Spray a coat of clear enamel gloss over the entire project. This is mostly for visual appeal but will also help prevent the thatch from shifting over time.

4. Drill two holes in the center back for mounting, 4 inches and 8 inches from the bottom. Insert deck screws to attach to post or tree. Before placing the box outside, open the trap door and put a few long, thin branches or a piece of bark inside to give butterflies a place to cling.

This house can be painted for aesthetic appeal before you spray on the enamel gloss.

thatched mailbox

This is the Night Mail crossing the Border
Bringing the cheque and the postal order.
Letters for the rich, letters for the poor,
The shop at the corner, the girl next door.
W. H. Auden

A great way to show off a thatched project is to place it in your front yard, and what better way than with a thatched-roof mailbox? This charming and functional project is sure to earn a few admiring comments from neighbors and postal workers alike. And though it might take a couple of tries to get a family of birds to nest in one of the birdhouse projects in this book, a mailbox will serve its purpose as soon as you place it outside.

Like the Tudor and Celtic cottage birdhouses, the mailbox allows for some creative expression. The location and shape of the decorative door and windows are guidelines. If you have a few scraps of wood lying around the workshop, this may be a perfect opportunity to use them.

Obviously not a birdhouse, this mailbox is a fun adaptation of the nuthatch Celtic cottage design.

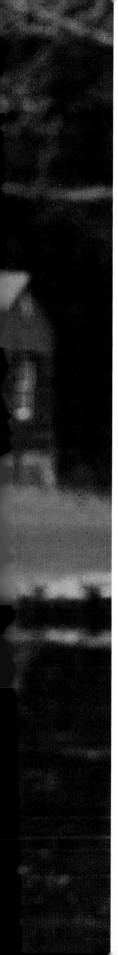

building a thatched mailbox

Difficulty: Moderate

This mailbox attaches to a wooden post. Check with your local postmaster for specific height and placement requirements. In general, mailboxes should be free of safety hazards (such as sharp edges or protruding nails), weathertight, and clearly visible. The mailbox door should open and close easily and should stay securely shut when closed. The flag should be easy for the mail carrier to see and should be a color that contrasts with the box. Check to see how close to the road you are allowed to place your mailbox.

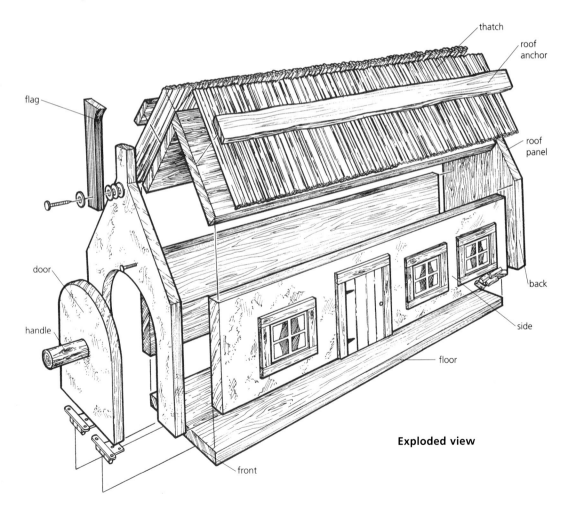

Exploded view

▶ what you'll need

▶ Parts

- Floor: ⅝" x 20" x 9" piece of plywood or solid pine
- 2 sides: ⅝" x 20" x 5" piece of plywood
- Front: ⅝" x 10" x 14" piece of plywood
- Back: ⅝" x 10" x 14" piece of plywood
- Roof panel: ⅝" x 20" x 6" piece of plywood
- Roof panel: ⅝" x 20" x 6¾" piece of plywood
- Door: ½" x 3¼" x 4½" piece of plywood
- Doorstop: scrap of solid wood about 1" x 1" x 2"
- 2 or 3 windows: ½" x 3½" x 2½" solid cedar, yellow pine, or maple
- Lintel: ½" x 4" x ¾" solid cedar, yellow pine, or maple
- Flag: ¾" x 1½" x 6" solid cedar, yellow pine, or maple
- Handle: 2" diameter x 1" scrap of wood for the handle (a piece of a branch or dowel works well)
- 4 roof anchors: ½" x 20" x 1" solid cedar, yellow pine, or maple
- 10 to 12 handfuls of thatch, 10" long
- Mounting board: ¾" x 16" x 6" piece of plywood
- Mounting post: 4 x 4 post, 7' long

▶ Materials

- Sandpaper
- Green, red, and black enamel spray paint
- Exterior wood glue
- ⅜"-diameter x 1" galvanized deck screws
- ⅜"-diameter x 2" galvanized deck screws
- 6d (6-penny) or 2" galvanized nails
- White exterior paint
- White, green, and black paint pens (available from most hobby, craft, and hardware stores)
- 1" deck screw to attach handle
- Spring-mounted cabinet hinges
- 4 or 5 washers, ¼" hole

▶ Tools

- Drill with a screwdriver bit and ⅛", ¼", and ¹⁵⁄₁₆" drill bits
- Jigsaw
- Hammer
- 2" paintbrush
- Utility knife
- Multipurpose scissors

Assembling the Mailbox

1. Choose the side of the floor that looks better, and sand it thoroughly. This side will be inside the mailbox. Paint the floor with a spray enamel in your choice of color.

Apply glue to the sides, and hold them with your hand in position against the floor. Then drill three holes in each side and drive 2-inch galvanized deck screws through them, fastening the sides to the floor. Make sure the screws are flush with the sides. ▼

2. On the front and back pieces, choose one of the shorter edges as the top, and make a mark halfway across it. Measure down 7 inches from the top, and make a mark on each edge. For the back, connect the marks to lay out the slope of the roof, and cut along the line. ▼

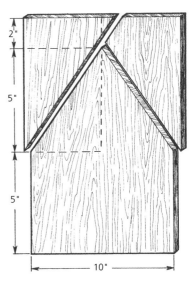

3. For the front, measure ½" to each side of the center top and connect these points to roof slope lines to form the chimney. Cut along the lines.

126 thatched mailbox

Making Mailbox Door and Doorstop

1. Create a paper template for the mailbox door using the measurements shown below. Then draw a line along the center of the front piece. Align the center of the template with this line, and align the bottom of the board with the bottom of the template.

Trace along the template to draw the arch at the top of the door. Then cut along the line with a jigsaw. ▼

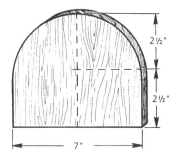

2. Attach the doorstop to the left inside of the door with glue and 1" nails, allowing a small lip to extend beyond the edge. This will keep the door from falling inside the box. With 6d nails, nail the front and back to the mailbox, lining up the side and bottom edges carefully. ▼

Painting and Decorating the Mailbox

1. Begin with the exterior, giving it a coat of exterior wood primer first and then two or three coats of white paint, until the wood grain is no longer visible.

2. While the paint on the box dries, prepare the decoration pieces, flag, and mailbox door. Begin with the decorative door. Use a utility knife to make three long score marks, evenly spaced, down the front. Spray the decorative door with red or green paint, covering it thoroughly. When it dries, draw a doorknob with a black paint pen. You may paint the lintel or leave it plain.

3. Cut pieces to size for the windows. Apply two coats of black paint. When the paint is dry, use a white paint pen to outline the "panes" and a green pen to draw curtains behind the panes. ▼

4. Sand the arch-shaped mailbox door, and paint it green.

Making the Flag

1. Make a paper template for the flag using the measurements and shape shown below. Trace template shape onto flag piece. Use jigsaw to cut out the shape. Sand edges if necessary and paint the flag bright red. ▼

Attaching the Roof, Decorative Details, and Door

1. Nail together the roof pieces, placing the side of the wider one over the edge of the narrower one to create a 90-degree angle. Put the roof between the front and back, and nail it into place. ▼

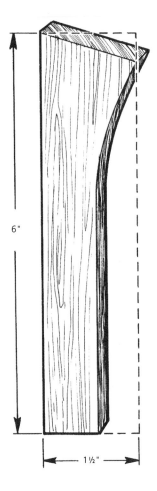

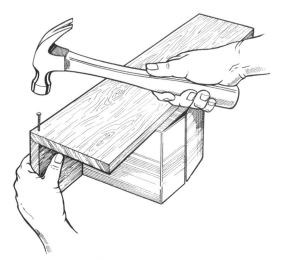

2. Attach the windows and decorative door to one side of the mailbox with exterior wood glue and ¾" nails. (If you like, you may also add bits of branch "logs" or other decorations.)

3. Drill a hole through the mailbox door from the back into the handle, and attach the handle using a 1-inch deck screw.

4. Use the spring hinges to attach the door to the box. Attach the shorter leaf of the hinge to the bottom of the door, so that the door will pull down and out as you pull it open. Rest the door in place, and attach the other end of the hinge to the bottom of the box. ▼

Affixing the Flag

1. The flag pivots on a nail driven in the front of the mailbox. Drill a ⅜-inch hole through the bottom end of the flag. Slide a washer onto a 2-inch galvanized nail, then slide the nail through the hole in the flag. Add about three or four more washers, which will keep the flag from rubbing against the mailbox. Drive the nail, with flag and washers attached, into the front of the mailbox near the top.

Drive another 2-inch nail into the front of the mailbox, as shown, to act as a stop for the flag when lowered. ▼

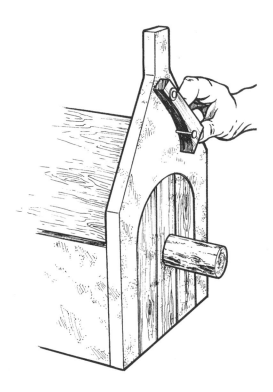

Thatching the Roof

1. Drill a series of five holes, evenly spaced, along the length of each of the 20-inch anchors. Place a 2-inch galvanized deck screw in each hole so that you will be able to secure the wood without having to reach for screws as you work.

2. Lean the mailbox against a piece of wood or another object so that one or the other side of the roof is level. Start placing small handfuls of thatch on it, patting them into place as you work. Apply enough to create a layer about 1 inch thick.

3. Place one of the wooden anchors across the thatch about 1 inch from the roof peak, and screw it loosely in place. The thatch should be secure, but still loose enough that you can slide it back and forth. Tap the top and bottom edges of the thatch lightly with a block to even out the edges, then attach a second anchor about 2 inches below the first one.

Thatch the second side of the roof the same way. Align the bottom edges of the thatch by tapping them with a wooden block, and then tap along the top of the roof to create an even peak. Once the thatch is even, tighten the screws in each anchor piece to secure firmly. Trim the top and bottom edges, using multipurpose scissors.

Mounting the Mailbox

1. The mailbox is attached to a mounting board, which, in turn, is attached to a post. Drill six ⅜-inch-diameter holes in the mounting board: one in each corner (½ inch from the edges) and two in the middle.

Check with the post office to see how far above the ground the box needs to be. (You'll need a post half again as long. Once you've attached the post to the box, bury the extra length in the ground, and pack soil firmly around it.) Place the mounting board on the post, and drill ¼-inch-diameter holes in the post, using the center holes in the mounting board as a guide. Drive ⅜-inch-diameter galvanized deck screws (the 2-inch ones) through the center holes to attach the mounting board to the top of the post.

Place the mailbox on top of the mounting board, and drill ¼-inch-diameter holes in the bottom of the mailbox, guided by the remaining holes in the mounting board. Drive ⅜-inch-diameter galvanized deck screws (the 1-inch ones this time) through the holes in the mounting board to attach it to the mailbox.

The post for your mailbox can be set close to the road for rural route delivery, or near your front steps for door-to-door foot delivery.

resources

Below is a list of resources and organizations that can provide you with more information on the topics discussed in this book.

The Backyard Wildlife Habitat Program offers information on attracting different animals to your yard and on the requirements for officially designating your yard a wildlife habitat.

The Backyard Wildlife Habitat Program
National Wildlife Federation
8925 Leesburg Pike
Vienna, VA 22184
phone: (703) 790-4000
Web site: www.nwf.org

For information on birds, bats, and butterflies in your in area or in general, take a look at the United States Fish and Wildlife Service Web site: www.fws.gov

The National Audubon Society also provides plenty of information on birds and other wildlife. Join the society, subscribe to *Audubon* magazine, see what's going on through their Web site (www.audubon.org) or write:

The National Audubon Society
700 Broadway
New York, NY 10003
phone: (212) 979-3000

General Bird Information

The Cornell Laboratory of Ornithology shares a wealth of bird knowledge, including backyard birding information, through their Cornell Nest Box Network. Check out the Web site: birds.cornell.edu

American Birding Association
P.O. Box 6599
Colorado Springs, CO 80934
phone: (719) 578-1614
fax: (719) 578-1480
Web site: www.americanbirding.org

National Bird Feeding Society
P.O. Box 23L
Northbrook, IL 60065-0023
phone: (847) 272-0135
fax: (847) 498-4092
e-mail: Feedbirds@aol.com
Web site: www.birdfeeding.org

Wild Birds Forever sells anything that the backyard birder needs, but it is also a source of information on birds in general.
27202 Highway 189
P.O. Box 4904
Blue Jay, CA 92317-4904
phone: (800) 459-BIRD
fax: (909) 336-6683
Web site: www.birdsforever.com/index.html

Bluebird Information

The North American Bluebird Society has a great Web site and is a resource for anyone interested in learning more about bluebirds.
The North American Bluebird Society
Dept. B
P.O. Box 74
Darlington, WI 53530-0074
Web site: www.nabluebirdsociety.org

For more information on bluebirds, including a personal perspective and links to other bluebird sites, check out www.nestbox.com/bluebird.htm

Owl Information

The Owl Pages
Web site: www.owlpages.com

Another Web site chock-full of owl information is www.rci.rutgers.edu/~au/owl.htm

Bat Sites

The best bat site, complete with info on bat houses, bat species, and its own bat magazine, belongs to Bat Conservation International. In addition to being a valuable site, it's just plain interesting.
Bat Conservation International
P.O. Box 162603
Austin, TX 78716
phone: (800) 538-BATS
Web site: www.batcon.org

Butterfly Sites

Butterflies of North America has an interactive map of North America showing what kind of butterflies live where.
http://www.npwrc.usgs.gov

The Butterfly Web Site
www.butterflywebsite.com

Thatching Information

For information about thatching, in addition to information on workshops and ways to obtain Norfolk reed, check out Colin's Web site at www.thatching.com/mn.html, e-mail Colin at Thatchit@aol.com, or call him toll-free at (888) THATCH-1 (888-842-8241).

metric conversion formulas

TO GET	WHEN YOU KNOW	MULTIPLY BY
centimeters	inches	2.54
meters	feet	0.305
meters	yards	0.9144

▶ other useful conversions

Kilograms from pounds, divide by 2.2

Degrees Celsius from degrees Farenheit, multiply by 5/9, then subtract 32

index

Page references in *italics* indicate illustrations; those in **bold** indicate charts.

Aligning thatch ends, 27, *27*, 28
America, thatching today in, 5–6
Amount of thatch, 14
Anchoring thatch, 14, 25, 27, *27*
Auden, W. H., 123

Bacon, Francis, 61
Bat Box, *60*, 64–69, *66–69*
Bats, 61–63, *63*
Bed, defined, 17
Belloc, Hilaire, 13
Big brown bat *(Eptesicus fuscus)*, 62–63, *63*
Bill-sweeping, 87
Bird Feeder, Thatched, *30*, 31–37, *33*, *35–37*
Birdhouses. *See* Projects
Birds. *See specific birds*
Birds, attracting, 8–11, *8–11*, **11**
Black-capped chickadee *(Poecile atricapillus)*, 39–41, *40*
Bluebird Box, *70*, 77, *77–83*, 79–83
Bluebirds
  eastern bluebird *(Sialia sialis)*, 71, 72, *72*, 75–76
  food for, **11**, 72, 73, 74, 75
  mountain bluebird *(Sialia currucoides)*, 71, 74, *74*, 75–76
  nesting habits of, 8, 72, 73, 74, 75–76
  western bluebird *(Sialia mexicana)*, 71, 73, *73*, 75–76
Blue jays, 10
Bottle, defined, 17
Brow course, defined, 17
Bunch/bundle, defined, 17
Butt, defined, 17
Butterflies, 113–15, *115*
Butterfly House, *112*, 116–21, *117–21*

Caching, 41
Cardinals, food for, 10, **11**
Celtic Cottage, Nuthatch, *84*, *91*, 91–97, *93–95*, 97
Chickadee Box, *38*, 39, *42*, 42–47, *44–47*
Chickadees
  black-capped chickadee *(Poecile atricapillus)*, 39–41, *40*
  food for, 9, 10, **11**, 40, 41
  nesting habits, 8, 40, 41
Clean-out doors, 46, *46*, 82, *82*
Clipping shears, 14
Competing birds, keeping away, 75
Course, defined, 17
Cross rods, defined, 17
Cutting thatch, 18

Decorating ideas, 69, *69*, 94, *94–95*, 111, 127, *127*
Drying thatch, 18
Dummy nests, 101

Eastern bluebird *(Sialia sialis)*, 71, 72, *72*, 75–76
Eastern screech owl *(Otus asio)*, 49–52, *50*
Edge strips, attaching, 23, *23*, 26, *26*
Egg dumping, 76
Ely Challenge Cup, 6
Eptesicus fuscus (big brown bat), 62–63, *63*
Equipment for thatching, 15–16, *15–16*, 19, *19*
European history of thatching, 2–4

Feeder for learning, *22*, 22–23. *See also* Projects
Fire danger of thatching, 3, 4
Food to attract birds, 9–11, **11**, *11*. *See also specific birds*

135

G

Gable flue, defined, 17
Gable top, defined, 17
Gardens for butterflies, 114
Goldfinches, food for, 10, **11**

H

Handful of thatch, measuring, 24
History of thatching, 2–4
House wren *(Troglodytes aedon)*, 99–101, *100*

K

Keats, John, 71

L

Lasiurus cinereus (little brown bat), 63, *63*
Learning to thatch, 21–29
    aligning thatch ends, 27, *27*, 28
    anchoring thatch, 14, 25, 27, *27*
    edge strips, attaching, 23, *23*, 26, *26*
    feeder selection for, *22*, 22–23
    handful of thatch, measuring, 24
    roof, *23*, 23–29, *26*, 26–29
    thatch preparation, 17, 24
    trimming thatch edges, 19, 28, *28*
Liggers, defined, 17
Little brown bat *(Lasiurus cinereus)*, 63, *63*
Long straw, 17

M

Mailbox, Thatched, *122*, 123–31, *124*, *126–31*
Master Thatcher, 6, *6*
Materials and techniques, 13–19. *See also specific projects*
    amount of thatch, 14
    anchoring thatch, 14, 25, 27, *27*
    clipping shears, 14
    cutting thatch, 18
    drying thatch, 18
    equipment, 15–16, *15–16*, 19, *19*
    long straw, 17
    safety tips, 16, *16*
    terminology, 17
    thatch preparation, 17, 24
    trimming thatch edges, 19, 28, *28*
    wood for building, 19, *19*
McGhee, Colin (Master Thatcher), 6, *6*, 133

Milton, John, 49
Mountain bluebird *(Sialia currucoides)*, 71, 74, *74*, 75–76

N

Nesting habits of birds, 8–9, *8–9*. *See also specific birds*
Norfolk reed, 2
Nuthatch Celtic Cottage, *84*, *91*, 91–97, *93–95*, *97*
Nuthatches
    food for, 9, 10, **11**, 86, 87, 89, 90
    nesting habits, 8, 86, 87, 89, 90
    red-breasted nuthatch *(Sitta canadensis)*, 85, *89*, 89–90
    white-breasted nuthatch *(Sitta carolinesis)*, 85, *86*, 86–88

O

Orioles, food for, **11**
Otus asio (eastern screech owl), 49–52, *50*
Otus kennicoottii (western screech owl), 49, *53*, 53–54
Owls. *See* Screech owls

P

Pinnacle, defined, 17
Poecile atricapillus (black-capped chickadee), 39–41, *40*
Projects
    Bat Box, *60*, 64–69, *66–69*
    Bluebird Box, *70*, 77, 77–83, *79–83*
    Butterfly House, *112*, 116–21, *117–21*
    Chickadee Box, *38*, 39, *42*, 42–47, *44–47*
    Nuthatch Celtic Cottage, *84*, *91*, 91–97, *93–95*, *97*
    Screech Owl House, *48*, *55*, 55–59, *57–59*
    Thatched Bird Feeder, *30*, 31–37, *33*, *35–37*
    Thatched Mailbox, *122*, 123–31, *124*, *126–31*
    Tudor Wren House, *98*, *102*, 102–11, *104–10*

R

Raccoon guard, 82, *82*
Rain resistance of thatch, 5
Red-breasted nuthatch *(Sitta canadensis)*, 85, *89*, 89–90
Reed for thatching, 2
Repair ease of thatching, 5
Ridge, defined, 17

Robins, food for, **11**
Roofs, thatched
　examples, *3–4, 6–7*
　instructions, *23,* 23–29, *26, 26–29*
　panels cutting tip, 35
　trimming top and bottom, 28, *28*
Rye straw, defined, 17

S afety tips, 16, *16*
Scatterhoarding, 87
Screech Owl House, *48, 55,* 55–59, *57–59*
Screech owls
　Eastern screech owl *(Otus asio),* 49–52, *50*
　food for, 50, 51, 53, 54
　nesting habits, 50, 52, 53, 54
　predators, 51
　western screech owl *(Otus kennicoottii),* 49, 53, *53*–54
Sialia currucoides (mountain bluebird), 71, 74, *74,* 75–76
Sialia mexicana (western bluebird), 71, 73, *73,* 75–76
Sialia sialis (Eastern bluebird), 71, 72, *72,* 75–76
Signing a roof, 7
Siskins, food for, 10, **11**
Sitta canadensis (red-breasted nuthatch), 85, *89,* 89–90
Sitta carolinesis (white-breasted nuthatch), 85, *86,* 86–88
Squirrels and bird feeders, 32
Straw for thatching, 2
Sweep, defined, 17

T ennyson, Lord Alfred, 39
Thatched Bird Feeder, *30,* 31–37, *33, 35–37*
Thatched Mailbox, *122,* 123–31, *124, 126–31*
Thatching. *See also* Learning to thatch; Materials and techniques; Projects
　in America today, 5–6
　benefits of, 5
　for bird houses, 7–11
　birds, attracting, 8–11, *8–11,* **11**
　defined, 2
　fire danger of, 3, 4

food for birds, 9–11, **11,** *11*
history in Europe, 2–4
Master Thatcher, 6, *6*
nesting habits of birds, 8–9, *8–9*
Norfolk reed, 2
rain resistance of, 5
reed for, 2
repair ease of, 5
resources for, 132–33
roofs made of, *3–4, 6–7*
signing a roof, 7
straw for, 2
water reed for, 2
yearns of straw, 2
Thatch preparation, 17, 24
Thermoregulation, 76
Thoreau, Henry David, 85
Trap doors, 58, *58, 119,* 119–20
Trimming
　roof, top and bottom of, 28, *28*
　thatch edges, 19
Troglodytes aedon (house wren), 99–101, *100*
Tse, Chuang, 113
Tudor Wren House, *98, 102,* 102–11, *104–10*
Tufted titmice, **11,** 76

W ater reed for thatching, 2
Waxwings, food for, **11**
Western bluebird *(Sialia mexicana),* 71, 73, *73,* 75–76
Western screech owl *(Otus kennicoottii),* 49, 53, *53*–54
White-breasted nuthatch *(Sitta carolinesis),* 85, *86,* 86–88
Whitman, Walt, 99
Wood for building, 19, *19*
Wordsworth, William, 31
Wotton, Sir Henry, 21
Wren House, Tudor, *98, 102,* 102–11, *104–10*
Wrens, **11,** 99–101, *100*

Y ealm, defined, 17
Yearns of straw, 2
Yeats, William Butler, 1

other storey titles you will enjoy

Hand-Feeding Backyard Birds, by Hugh Wiberg. With simple instructions and over 80 color photos, this book explains how to get up-close and personal with your feathered friends. Learn which birds are most receptive to hand-feeding; their favorite foods; and the best times, places, and weather conditions for hand-feeding. Tells how to capture amazing photographs of birds on the hand. 160 pages. Paperback. ISBN 1-58017-181-8.

The Backyard Birdhouse Book: Building Nestboxes and Creating Natural Habitats, by René and Christyna M. Laubach. Written by two experienced naturalists, this practical guide offers complete plans for 8 easy-to-build birdhouses tailored to the needs of 25 cavity-nesting species. Also includes information on monitoring bird activity and protecting birds from predators, parasites, and competing species. Full-color photographs and illustrations throughout. 216 pages. Paperback/hardcover. ISBN 1-58017-104-4/1-58017-172-9.

Everything You Never Learned About BIRDS, by Rebecca Rupp. Intended to introduce children to the world of birding, this book overflows with a mazing facts, fun projects, and fascinating legends. 144 pages. Paperback. ISBN 0-88266-345-3.

The Backyard Bird-Lover's Guide, by Jan Mahnken. covers feeding, territory, courtship, nesting, laying, and parenting characteristics of many birds. Identification section describes 135 species and includes watercolor paintings. 320 pages. Paperback. ISBN 0-88266-927-3.

Birdfeeders, Shelters & Baths, by Edward A. Baldwin. Designs for a wide range of bird feeders and baths that will attract birds to your yard all year. 128 pages. Paperback. ISBN 0-88266-623-1.

Birdhouses: 20 Unique Woodworking Projects for Houses and Feeders, by Mark Ramuz and Frank Delicata. Designs range from classic and simple to whimsical and ornate. Includes plans with easy-to-follow illustrations, instructions, materials and tools lists, and ideas for how to finish, decorate, and weatherproof the birdhouse. 128 pages. Paperback. ISBN 0-88266-917-6.

Gifts for Bird Lovers, by Althea Sexton. Any bird fancier will be thrilled to make or receive these gifts, many of which are seen in popular catalogs and magazines. 128 pages. Paperback. ISBN 0-88266-981-8.

These books and other Storey Books are available at your bookstore, farm store, or garden center, or directly from Storey Books, 210 MASS MoCA Way, North Adams, MA 01247, or by calling 1-800-441-5700. Or visit our Web site at www.storey.com.